Building Resilience to Climate Change in South Caucasus Agriculture

DIRECTIONS IN DEVELOPMENT
Agriculture and Rural Development

Building Resilience to Climate Change in South Caucasus Agriculture

Nicolas Ahouissoussi, James E. Neumann, and
Jitendra P. Srivastava, Editors

THE WORLD BANK
Washington, D.C.

ISBN (paper): 978-1-4648-0214-0
ISBN (electronic): 978-1-4648-0215-7
DOI: 10.1596/978-1-4648-0214-0

Cover photo: © Asim Talib. Used with permission. Further permission required for reuse.
Cover design: Debra Naylor, Naylor Design

Library of Congress Cataloging-in-Publication Data

Building resilience to climate change in South Caucasus agriculture / Nicolas Ahouissoussi, James E. Neumann, and Jitendra P. Srivastava, editors.
     pages cm. — (Directions in development)
     Includes bibliographical references.
ISBN 978-1-4648-0214-0 (alk. paper) — ISBN 978-1-4648-0215-7
    1. Climatic changes—Caucasus, South. 2. Resilience (Ecology)—Caucasus, South. I. Ahouissoussi, Nicolas. II. Neumann, James E., 1962- III. Srivastava, Jitendra, 1940- IV. World Bank. V. Series: Directions in development (Washington, D.C.)
    QC903.2.C28B85 2014
    630.2'086—dc23
                             2014008437

Contents

Figures

Maps

Tables

Foreword

Climate action is a key priority of the World Bank Group and a pillar of the Europe and Central Asia regional strategy. Improving the understanding of risks and assisting our clients in implementing adaptation and mitigation measures for a changing climate is an integral part of our development agenda. As World Bank Group President Jim Yong Kim noted at the opening of the Green Climate Fund: "We know that climate change in our lifetime threatens to roll back so many of the gains that we have made over the years...And we are convinced that there is no way that we will be able to end poverty by 2030, which is our goal—without tackling climate change in the most serious manner... Also we know that we need to feed 9 billion people by 2050, and the only way that we can do that is to make agriculture more resilient, more productive, in the changing landscapes that we will see due to climate change."

This book, *Building Resilience to Climate Change in South Caucasus Agriculture*, responds directly to the urgent need for climate adaptation, as highlighted in the World Bank's "Turn Down the Heat: Why a 4°C Warmer World Must Be Avoided" report. It is an example of the World Bank's commitment to helping countries respond to the opportunities and challenges posed by climate change.

The South Caucasus Region is already experiencing increasing aridity and more frequent extreme weather events (e.g., severe droughts, floods, and hailstorms). This book is a synthesis of country studies undertaken in collaboration with policy makers, farmers, civil society, and other stakeholders in Armenia, Azerbaijan, and Georgia. It provides practical pointers for action by quantifying the impact and identifying key priorities for policies, programs, and investments to reduce the vulnerability of agriculture to climate change. The book also presents practical solutions for a more climate-smart agriculture at the national level and highlights the potential benefits of greater regional collaboration.

We view this work as an important beginning. To achieve the goal of climate resilience in the agriculture sector, additional effort will be needed to translate the proposed solutions into reality. The analysis demonstrates that investments in irrigation infrastructure and improved on-farm technologies have great potential to raise agricultural productivity and improve the climate resilience of the sector. Demand-side agricultural water management will have high short-term payoffs, which will contribute to the success of long-term irrigation, drainage, and other agriculture and infrastructure investments. There is also a need to improve agriculture risk management strategies to help mitigate household risks from extreme events, especially for the poorest farmers, who are the most vulnerable.

We need to take action now to safeguard the gains that have been achieved in the agriculture and rural development sectors from the risks of climate change and to sustain them over the longer term. The World Bank stands ready to be an active partner in this agenda.

Laura Tuck   Rachel Kyte
Vice President   *Group Vice President and Special Envoy*
Europe and Central Asia Region   *Climate Change*
World Bank Group   *World Bank Group*

Preface

This volume is a synthesis of a multicountry collaborative effort between the World Bank and the governments of Armenia, Azerbaijan, and Georgia. This effort built on the success of a similar effort in four Eastern Europe and Central Asia (ECA) countries, detailed in the World Bank book, *Looking Beyond the Horizon* (Sutton, Srivastava, and Neumann, eds. 2013a). The goal of this new book in this second set of countries is to bring together the lessons learned and recommendations from the country-specific work, identify new insights for adaptation planning at the regional level for the South Caucasus, and provide further guidance on the approach and methodology for others who wish to pursue similar analyses elsewhere.

The effects of changes in climate on agricultural systems and rural economies are already evident throughout Europe and Central Asia. Adaptation measures now in use in the region, largely piecemeal efforts, will be insufficient to prevent the negative effects of climate change on agricultural production over the coming decades. Interest is growing at the regional, country, and development partner levels to gain a better understanding of the exposure, sensitivities, and impacts of climate change at the farm level, as well as to develop and prioritize adaptation measures to mitigate the adverse consequences.

Building from the findings and recommendations of the landmark report, *Adapting to Climate Change in Europe and Central Asia* (World Bank 2009), the World Bank in 2009 embarked on a program for select ECA client countries called the Regional Analytical and Advisory Activities Program on Reducing Vulnerability to Climate Change in ECA Agricultural Systems. Its purpose is to enhance the client countries' ability to mainstream climate change adaptation into agricultural policies, programs, and investments. This multistage effort has included activities to raise awareness of the threat, analyze potential impacts and adaptation responses, and build capacity among client country stakeholders and ECA Bank staff with respect to climate change and the agriculture sector.

The present study is the culmination of the efforts by national and regional institutions and researchers, the World Bank team, and a team of international experts headed by the consulting firm Industrial Economics, Incorporated, to analyze the potential impacts climate change may have on the agriculture sector in the client countries, but, more importantly, to develop a list of prioritized measures those countries can use to adapt to those impacts.

This volume details the results of this work, starting with the country-level results, then adding a unique regional perspective on the task of building resilience to climate change in the agriculture sector. The underlying country-level work identified a menu of climate change adaptation options for the agriculture and water resources sectors, along with specific recommended adaptation actions, tailored to distinct agricultural zones within the region. The detailed country reports for Armenia, Azerbaijan, and Georgia are available (Ahouissoussi et al. 2014a, 2014b, 2014c). The regional level results add the dimensions of shared water, climate, and ecological resources that present both new challenges and exciting opportunities for building resilience.

Early elements of the World Bank's program, beginning in 2009 in Albania, the former Yugoslav Republic of Macedonia, Moldova, and Uzbekistan, established the basic analytic framework while revealing several key lessons. The current program for the three South Caucasus countries of Armenia, Azerbaijan, and Georgia, builds significantly on the prior work in several key respects. First, the proximity of the countries allowed for focusing on the issues of transboundary water. Second, the countries were engaged in developing a unified set of climate scenarios for the analysis, facilitating better cross-country dialogue on adaptation. Third, a regional workshop conducted at the end of the study focused specifically on identifying opportunities for cooperation to rationalize the respective roles of the three countries. Such cooperation would take advantage of the respective economies of scale in orienting research and agricultural extension toward identified climate change challenges, and it would begin a process for sharing results of efforts to build a strategic action plan for climate change adaptation.

This work is also timely in directly responding to the urgent needs for climate adaptation established in the World Bank's recent series of "Turn Down the Heat" reports (World Bank 2013b), with focus on the ECA region where many countries struggle with increased aridity, more frequent climate extremes (for example, severe and prevalent droughts and hailstorms), and the challenges of reforming the agricultural economy, land tenure, and land use, while maximizing the utility of the former Soviet irrigation systems in place in all three countries. This study demonstrates the next step in the adaptation process for these countries, as well as countries in similar regions that face a global problem: providing actionable recommendations at the country and subnational-region scales, which have been developed with the comprehensive and active involvement of local stakeholders ranging from policy makers to farmers. Further, the study has underlined the importance and urgency of capacity-building, empowering these countries to initiate control of their own climate resilience while also providing specific guidance to finance opportunities in the rapidly emerging climate adaptation sector. With the risks of inaction clarified and adaptation options identified and agreed upon, the path ahead can focus on taking action.

This book summarizes the findings of the country reports for these three South Caucasus countries and provides a broader harmonized regional perspective. It identifies the regional elements of climate and ecosystems important to the success of the agriculture sector, including land, water, and vegetation.

The detailed country-level analytic work, various stakeholder consultations, and consensus-building work are at the heart of the effort. As such, this study takes steps to synthesize some common regional-scale adaptation measures for the forecasted hotter and drier climate where the key similarities and differences among the three countries' agreed adaptation plans are also identified. The result is a list of specific actions that any of the three countries could consider exploiting as regional-scale opportunities in building resilience to climate change.

References

Ahouissoussi, N., J. E. Neumann, J. P. Srivastava, B. B. Boehlert, and S. Sharrow. 2014a. *Reducing the Vulnerability of Armenia's Agricultural Systems to Climate Change: Impact Assessment and Adaptation Options.* Washington, DC: World Bank.

Ahouissoussi, N., J. E. Neumann, J. P. Srivastava, C. Okan, B. B. Boehlert, and K. M. Strzępek. 2014b. *Reducing the Vulnerability of Azerbaijan's Agricultural Systems to Climate Change: Impact Assessment and Adaptation Options.* Washington, DC: World Bank.

Ahouissoussi, N., J. E. Neumann, J. P. Srivastava, C. Okan, and P. Droogers. 2014c. *Reducing the Vulnerability of Georgia's Agricultural Systems to Climate Change: Impact Assessment and Adaptation Options.* Washington, DC: World Bank.

World Bank. 2009. *Adapting to Climate Change in Europe and Central Asia.* Washington, DC: World Bank (accessed October 7, 2013), http://www.worldbank.org/eca /climate/ECA_CCA_Full_Report.pdf.

———. 2013a. *Looking Beyond the Horizon: How Climate Change Impacts and Adaptation Responses will Reshape Agriculture in Eastern Europe and Central Asia*, edited by W. R. Sutton, J. P. Srivastava, and J. E. Neumann. Report 76184, Washington, DC: World Bank.

———. 2013b. "Turn Down the Heat: Climate Extremes, Regional Impacts, and the Case for Resilience." Report, World Bank, Washington, DC (accessed December 9, 2013), http://www.worldbank.org/en/topic/climatechange/publication/turn-down-the-heat -climate-extremes-regional-impacts-resilience.

Acknowledgments

This book was prepared by a team led by Nicolas Ahouissoussi of the World Bank Sustainable Development Department, Europe and Central Asia Region, together with Nedret Durutan, Cüneyt Okan, Jitendra Srivastava, Darejan Kapanadze, Rufiz Vakhid Chirag-Zade, and Arusyak Alaverdyan, in collaboration with Industrial Economics, Incorporated. The Industrial Economics team was led by James Neumann. We are grateful to Dina Umali-Deininger, Sector Manager, Agriculture and Rural Development, Sustainable Development Department, Europe and Central Asia Region, for valuable support and guidance. We are also grateful to Henry Kerali, Country Director, South Caucasus Country Unit based in Georgia, for his support in furthering the agenda on climate change in agriculture, and to Larisa Leshchenko, Azerbaijan Country Manager, and Jean-Michel Happi, Armenia Country Manager. We also gratefully acknowledge Larysa Hrebianchuk for providing administrative support.

The drafting of this book is the result of a collaborative effort of a large number of individuals and organizations. The authors of each chapter are as follows:

Chapter 1—Introduction and Reasons for Action: Nicolas Ahouissoussi, James Neumann, Jitendra Srivastava, and independent consultants, Nedret Durutan and Cüneyt Okan.

Chapter 2—Framework and Program Design: Nicolas Ahouissoussi, James Neumann, Jitendra Srivastava, Kenneth Strzępek (Massachusetts Institute of Technology), Peter Droogers (FutureWater), Stephen Sharrow (Oregon State University), and Brent Boehlert (Industrial Economics, Inc.).

Chapters 3 through 5—Risks, Impacts, and Adaptation Menus for Study Countries: Brent Boehlert, James Neumann, Kenneth Strzępek, Peter Droogers, and Stephen Sharrow.

Chapter 6—Climate Change Impacts and Adaptation Options in the South Caucasus Region: Nicolas Ahouissoussi, James Neumann, Kenneth Strzępek, and Brent Boehlert.

Chapter 7—Adaptation in the South Caucasus: Opportunities for a Regional Approach: Nicolas Ahouissoussi, James Neumann, Jitendra Srivastava, Nedret Durutan, and Cüneyt Okan.

Other contributors from Industrial Economics, Incorporated, include Margaret Black, Ellen Fitzgerald, Miriam Fuchs, and Nicholas Tyack. World Bank reviewers included Ana Bucher, Stephen Mink, and Kanta Rigaud. Channing Arndt of United Nations University, World Institute of Development Economics Research, provided external peer review.

From the governments of Armenia, Azerbaijan, and Georgia, we are grateful for policy guidance and support provided by the Ministries of Agriculture and Environment, the three country steering committees, and the study focal points, including Medea Inashvili and Konstantine Kobakhidze for Georgia; Armen Poghosian for Armenia; and Yolchu Zeynalov, Ogtay Jafarov, and Mehriban Kazimova for Azerbaijan, without whom this study would not have been possible. The study greatly benefited from the valuable inputs, comments, advice, and support provided by academia, civil society, and nongovernmental organizations farmers, the donor community, and development partners in all three countries throughout this work.

The funding for this study by the Bank-Netherlands Partnership Program is gratefully acknowledged.

About the Editors

Nicolas Ahouissoussi is Senior Agriculture Economist in the World Bank's Europe and Central Asia Region, Agricultural and Rural Development Unit. Prior to joining the ECA Region, he was Senior Agriculture Economist in the World Bank Africa Region. He has over 30 years of work experience in the economic and agriculture sectors, 17 of which were for the World Bank. He holds a PhD in Agricultural and Applied Economics from the University of Georgia, USA.

James E. Neumann is Principal and Environmental Economist at Industrial Economics, Incorporated, a Cambridge, Massachusetts, consulting firm that specializes in the economic analysis of environmental policies. Neumann is the coeditor with Robert Mendelsohn of *The Impact of Climate Change on the United States Economy*, an integrated analysis of economic welfare impacts in multiple economic sectors, including agriculture, water resources, and forestry. He specializes in the economics of adaptation to climate change and is a lead author for the Intergovernmental Panel on Climate Change Working Group II chapter on the "Economics of Adaptation."

Jitendra P. Srivastava, former Lead Agriculturist at the World Bank, is globally recognized for his contributions in the fields of agricultural research, education, agri-environmental issues, and the seeds sector. Prior to working at the World Bank, he served in leadership and technical roles at the International Center for Agricultural Research in the Dry Areas, the Ford Foundation, and the Rockefeller Foundation, and was professor of genetics and plant breeding at Pantnagar University, India, where he received the first Borlaug Award for his contribution to the Indian Green Revolution. He holds a PhD from the University of Saskatchewan, Canada, in plant genetics. Srivastava is a fellow of several national academies of sciences and is the recipient of honorary doctorates from four agricultural universities.

Abbreviations

B-C	benefit-cost
BFI	bilateral financial institution
BNPP	Bank-Netherlands Partnership Program
BWO	basin water organization
CA	conservation agriculture
CLIRUN	Climate and Runoff Hydrologic Model
CGIAR	Consultative Group on International Agricultural Research
CIMMYT	International Maize and Wheat Improvement Center
CMI	climate moisture index
ECA	Europe and Central Asia
EECCA	Eastern Europe, Caucasus, and Central Asia
EU	European Union
FAO	Food and Agriculture Organization of the United Nations
FPU	food production unit
GCM	General Circulation Model
GDP	gross domestic product
GHG	greenhouse gas
ha	hectare
ICARDA	International Center for Agricultural Research in the Dry Areas
ICWC	Interstate Commission for Water Coordination
IFAD	International Fund for Agricultural Development
IFI	international financial institution
IFPRI	International Food Policy Research Institute
IPCC	Intergovernmental Panel on Climate Change
KGCCS	Köppen-Geiger Climate Classification System
km	kilometers
MCM	million cubic meters
mm	millimeter
NBFI	non-bank financial institutions

NGO	nongovernmental organization
NPV	net present value
O&M	operations and maintenance
PET	potential evapotranspiration
SEI	Stockholm Environment Institute
SRES	IPCC *Special Report on Emissions Scenarios*
UNDP	United Nations Development Programme
UNFCCC	United Nations Framework Convention on Climate Change
USAID	United States Agency for International Development
WEAP	Water Evaluation and Planning System
WFD	Water Framework Directive of the European Union
WMO	World Meteorological Organization
WWF	World Wildlife Fund

Overview

Agricultural production is inextricably tied to climate, making agriculture one of the most climate-sensitive of all economic sectors. In countries like Armenia, Azerbaijan, and Georgia, the risks of climate change for the agriculture sector are a particularly immediate and important problem because the majority of the rural population depends either directly or indirectly on agriculture for their livelihoods. Climate change disproportionately affects the rural poor because of their greater dependence on agriculture, their relatively lower ability to adapt, and the high share of income they spend on food. Climate change effects could therefore undermine progress that has been made in poverty reduction and adversely impact food security and economic growth in vulnerable rural areas.

International efforts to limit greenhouse gases (GHGs) and to mitigate climate change are urgently needed to prevent the adverse effects of temperature increases, changes in precipitation, and the increased frequency and severity of extreme weather events. At the same time, climate change can also create economic opportunities, particularly in the agriculture sector. Increased temperatures can lengthen growing seasons, higher carbon dioxide concentrations can enhance plant growth, and in some areas rainfall and the availability of water resources can increase as a result of climate change.

If countries are to effectively manage the risks of climate change—and take advantage of potential opportunities—it is necessary to develop a clear plan for aligning agricultural policies with climate change, for developing key agricultural institutions' capabilities, and for making needed infrastructure and on-farm investments. Developing such a "climate-smart" plan ideally involves access to a combination of high-quality quantitative analysis and consultations with key stakeholders, particularly farmers, as well as local agricultural experts—and these analyses and consultations must take explicit account of the uncertainty of future climate as well. The most effective plans for adapting this sector to climate change will involve both human and physical capital enhancements, many of which may also enhance agricultural productivity under current climate variability and conditions, giving the measures a "win-win" quality. In the South Caucasus countries, climate-smart plans are focused on adapting the sector to climate change as a primary goal, but many of the "win-win" opportunities can also achieve GHG emission reduction goals for the sector, thus making them "win-win-win" options.

The country-level efforts presented in this study have identified a set of measures at the national and agricultural region levels that have the potential to dramatically increase the resilience of the region's agriculture to climate change, while simultaneously reducing the GHG emissions footprint of the sector. The measures identified at the country and subnational region scales are very similar across the region, suggesting that a regional approach for adaptation may provide economies of scale, save precious resources, and enhance opportunities to more effectively manage shared water resources compared to independent and decoupled national efforts from each of the three countries.

Putting these plans for increasing resilience into action requires concerted effort at the national and regional levels to the extent practicable. This study is designed to facilitate action at the regional level for the South Caucasus countries by: (1) sharing relevant information on the current scientific, economic, and policy context that motivates action; (2) providing an analytic basis for prioritizing measures at the country and regional levels that recognizes the uncertainty of future climate outcomes; and (3) recommending a series of actions at the regional level. The study acknowledges the complex physical links and political dynamics of the region, while also identifying key linkages and synergies in taking the "regional" approach.

Key Findings

The study's key findings fall into two general categories: (1) exposure of agricultural systems to climate change, particularly changes in temperature and precipitation, with resulting effects on crop yields and (2) adaptive capacity of agricultural systems, given the national socioeconomic, technical, and institutional contexts, which leads to recommendations for location-specific adaptation options on the basis of both quantitative and qualitative analysis.

Projecting the Exposure of Agricultural Systems to Climate Change

The study was conducted in the three countries of the South Caucasus region—Armenia, Azerbaijan, and Georgia—at both the country and agricultural region levels. Each of the subnational agricultural regions (map O.1) not only exhibits similar characteristics within its boundaries in terms of terrain, climate, soil type, and water availability, but also differs from neighboring regions in ways that are important to adaptation planning.

The study scope included the crop, livestock, and irrigation sectors—that is, systems within the managed agriculture sector. Baseline agricultural conditions, probable climate change impacts, and available adaptive options were found to be similar within each of the regions in map O.1 in ways that are significant for developing a specific "adaptation plan." On the map, darker areas are high elevation (typically characterized by mixed livestock/cereal with some high-value fruit production), and lighter areas are low elevation (typically characterized by irrigated production of high-value vegetables and fewer cereals and, in the case

Map O.1 Agricultural Regions of the South Caucasus

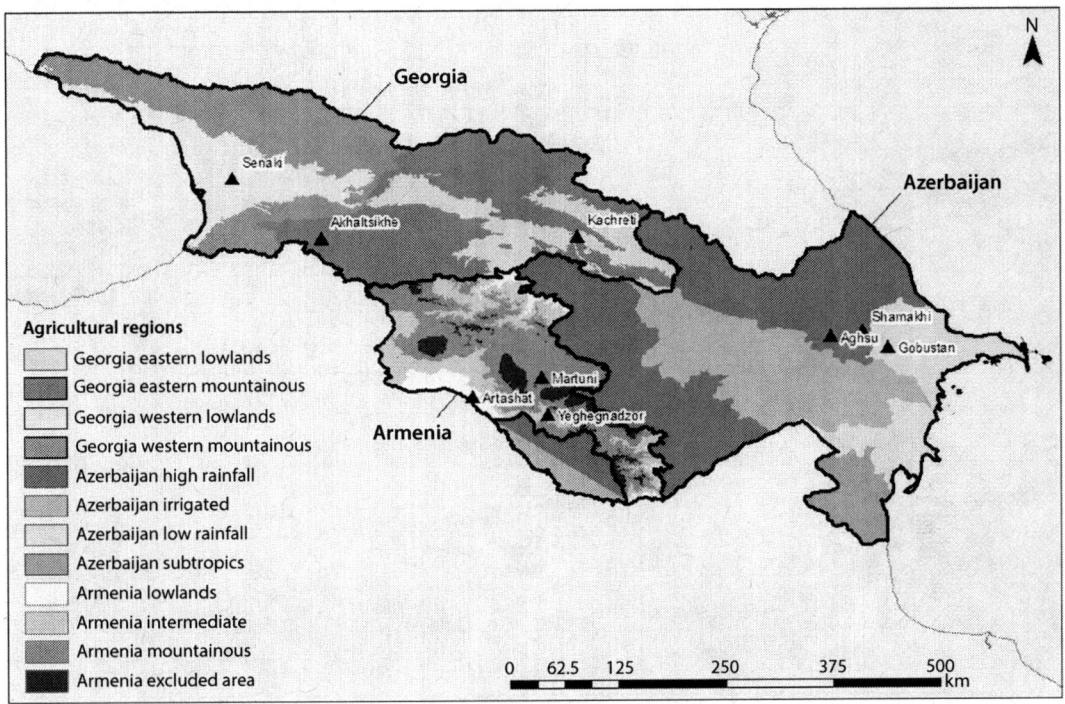

Sources: ©Industrial Economics. Used with permission; reuse allowed via Creative Commons Attribution 3.0 Unported license (CC BY 3.0). Country boundaries are from ESRI and used via CC BY 3.0.
Note: km = kilometers.

of Azerbaijan, potential for cotton production). Contiguous areas of high elevation are common throughout the region.

Three climate change scenarios were employed in the study: (1) Low Impact, (2) Medium Impact, and (3) High Impact. The study included results in each country for these three climate scenarios because of the uncertainty in these climate forecasts. Precipitation forecasts are particularly uncertain, both in terms of the magnitude and the direction of change—that is, whether precipitation is likely to increase or decrease under climate change. However, a key finding of the study is that the adaptation measures would be robust responses to climate change under all three climate scenarios.

Map O.2 shows the effect of climate change on annual average temperature and precipitation under the study's Medium Impact Scenario. The map shows temperatures increasing across the region and precipitation increasing in some areas but decreasing in others. In areas where temperature increases and precipitation decreases, the resulting aridity and reduced soil moisture would likely have negative consequences for agriculture.

The yearly averages shown in map O.2 are less important for agricultural production than the seasonal distribution of temperature and precipitation. Forecasted temperature increases are highest in September and precipitation decreases are greatest in July and August relative to current conditions, as

Map O.2 Effect of Climate Change on Annual Average Temperature and Precipitation in the 2040s under Medium Impact Climate Scenario

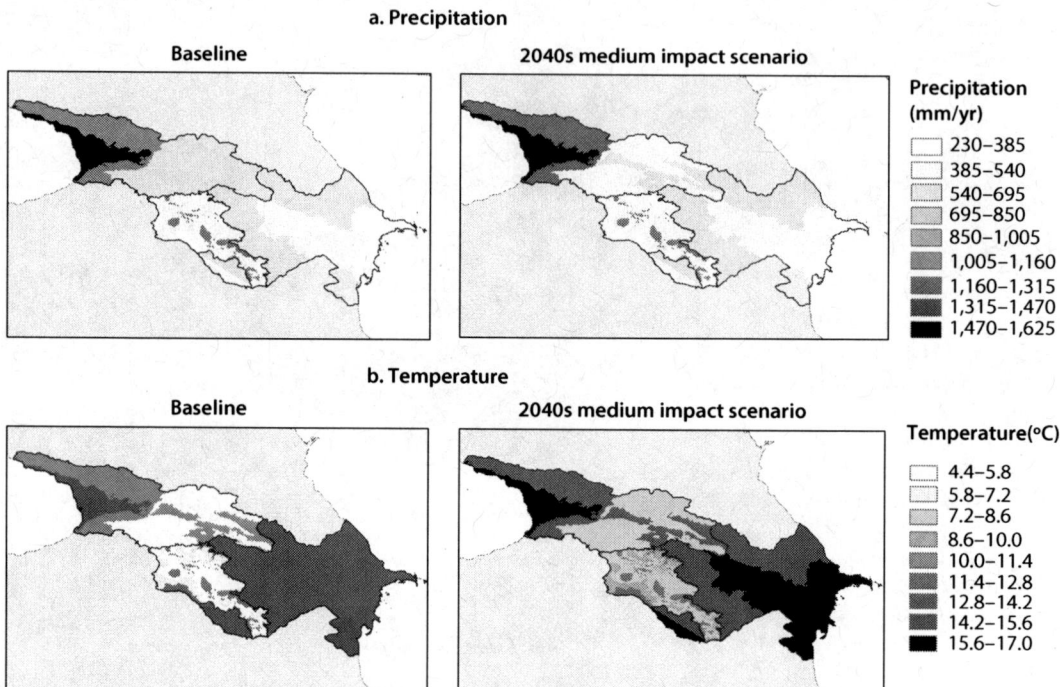

a. Precipitation

Baseline   2040s medium impact scenario

Precipitation (mm/yr)
- 230–385
- 385–540
- 540–695
- 695–850
- 850–1,005
- 1,005–1,160
- 1,160–1,315
- 1,315–1,470
- 1,470–1,625

b. Temperature

Baseline   2040s medium impact scenario

Temperature(°C)
- 4.4–5.8
- 5.8–7.2
- 7.2–8.6
- 8.6–10.0
- 10.0–11.4
- 11.4–12.8
- 12.8–14.2
- 14.2–15.6
- 15.6–17.0

Sources: ©Industrial Economics. Used with permission; reuse allowed via Creative Commons Attribution 3.0 Unported license (CC BY 3.0). Country boundaries are from ESRI and used via CC BY 3.0.
Note: mm/yr = millimeters per year.

illustrated in figure O.1 for Georgia. This September temperature increase can be as much as 5°C in lower elevation agricultural regions, when temperatures are already near their highest. In addition, forecasted precipitation declines are greatest in the agriculturally critical May-to-October period, causing the late summer and early fall to be the driest times of year under all reviewed climate scenarios.

If no adaptation actions are taken, the impact of these changes on crop yields could be severe. The impact of climate change in the Medium Impact Scenario may reduce yields of rainfed crops by 3–28 percent through 2050. Irrigated crops could see more modest yield reductions, of 3–16 percent, but only if sufficient irrigation water will be available.

For irrigated crops a critical factor is whether there will be sufficient water under a changed climate to maintain irrigation. With increased temperatures, crops will require more irrigation to maintain the yields achieved today. In addition, higher temperatures can reduce water runoff into rivers, so less water is available in the rivers for irrigation. Analyses conducted for the study demonstrate that irrigation water shortages can be expected in six basins of the region, even without climate change, because water is removed for other uses, such as hydropower and municipal and industrial water supplies, before reaching the irrigated areas. These shortages, however, would be much more severe under climate change.

Figure O.1 Effect of Climate Change on Monthly Temperature and Precipitation Patterns by 2040s for Georgia's Eastern Lowlands Agricultural Region

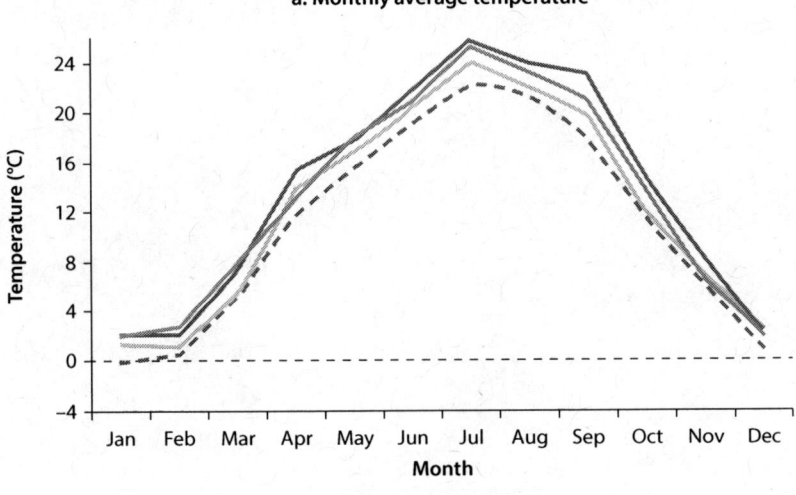

a. Monthly average temperature

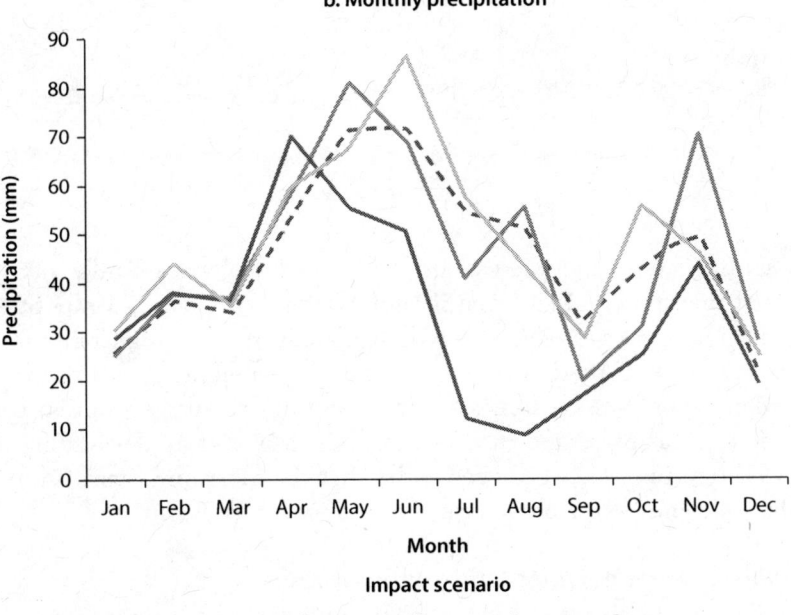

b. Monthly precipitation

Impact scenario

━━ High ━━ Medium ⋯⋯ Low ━ ━ Base

Source: World Bank data.
Note: mm = millimeter.

The six basins shown in map O.3 are those forecasted to have irrigation water shortages in the 2040s under all climate scenarios. They include some of the most productive, high-value fruit and vegetable production areas in Azerbaijan and Armenia, as well as the area of some of the best wine grapes in Georgia.

If farmers do not have sufficient water available for irrigation, they must either reduce their cultivated area or suffer reduced yields compared to

Building Resilience to Climate Change in South Caucasus Agriculture
http://dx.doi.org/10.1596/978-1-4648-0214-0

Map O.3 Basins with Forecasted Irrigation Water Shortages by 2050, under all climate change scenarios

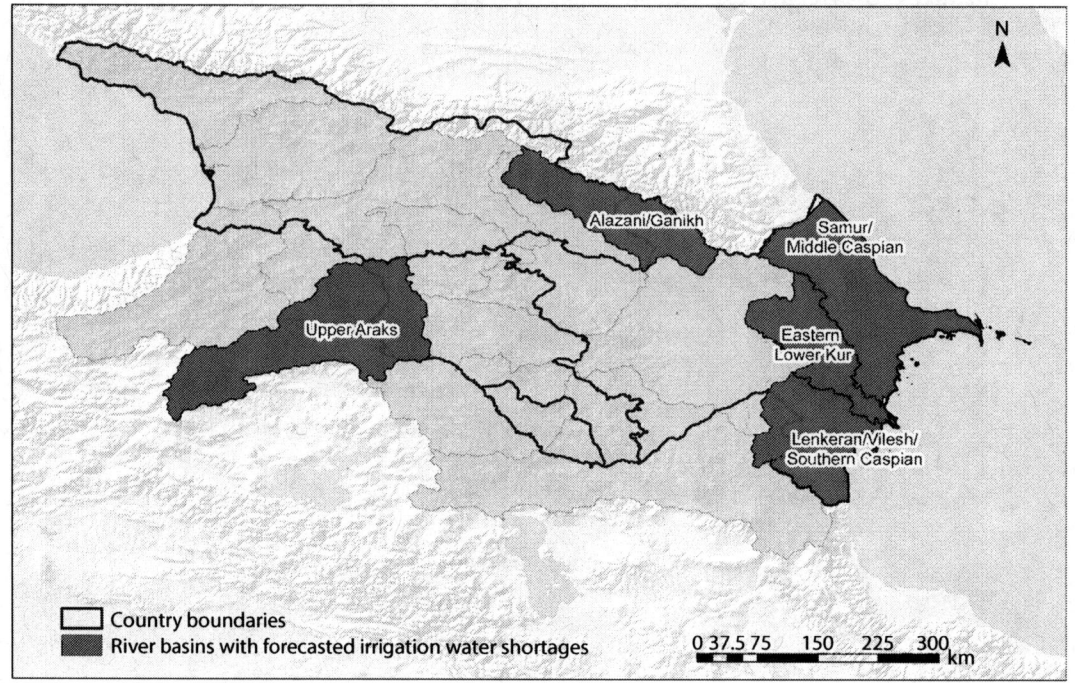

the potential if and when the water demands of the plants are fully met. A key finding of this study is that, even if climate does not change, competition for water from nonagriculture sectors will likely reduce water availability for agriculture. Yet, when climate change is taken into account, irrigated crop yields in areas where shortages are forecasted are reduced on average from 30 percent to 77 percent, as presented in table O.1, which would be devastating to the region's agriculture. The study concludes that the most important risk to agricultural yields in the region is water availability for irrigation.

National and Regional Adaptation Approaches

The study team developed an extensive list of potential adaptation options that might be considered to reduce risks posed by climate change to crops and livestock in each country. Measures for adoption at the national level were identified based on quantitative and qualitative analysis of potential net benefits (including quantitative benefit-cost analyses), as well as evaluations and recommendations from farmer stakeholders and expert groups. Each adaptation option assessed is supported by a graph that shows benefit-cost (B-C) ratios for each crop for the baseline and each climate scenario and under two price scenarios.

Stakeholders then participated in small groups to consider proposed adaptation measures at a series of National Dissemination and Consensus-Building

Conferences held in each country's capital city. For each measure, the groups considered (1) B-C analysis results (net economic benefit), (2) the potential of an option to increase agricultural productivity with or without climate change, and (3) the GHG mitigation potential.

There was significant overlap among the recommended national and regional adaptation options for Armenia, Azerbaijan, and Georgia, as shown in figures O.2 and O.3. Five of the seven national climate change adaptation

Table O.1 Effect of Climate Change on Irrigated Crop Yields in the 2040s Relative to Current Yields under the Medium Impact Climate Scenario, Adjusted for Estimated Irrigation Water Deficits

	Agricultural region (country)/river basin, percentage change in yield		
Crop	E. lowlands (GE) Alazani	Irrigated (AZ) S. Caspian	Lowlands (AR) Upper Araks
Alfalfa	n.a.	–77	–48
Corn	–33	–77	n.a.
Grape	–30	–66	–42
Potato	–34	–77	–51
Tomato	–35	n.a.	–53
Wheat	–34	–77	–48

Source: World Bank data.
Notes: Results are average percentage change in crop yield, assuming no effect of carbon dioxide fertilization. Declines in yield are shown in shades of orange, with darkest representing biggest declines. AR = Armenia; AZ = Azerbaijan; GE = Georgia; n.a. = not applicable (indicates that the crop was not analyzed in that country).

Figure O.2 National-Level Recommended Measures

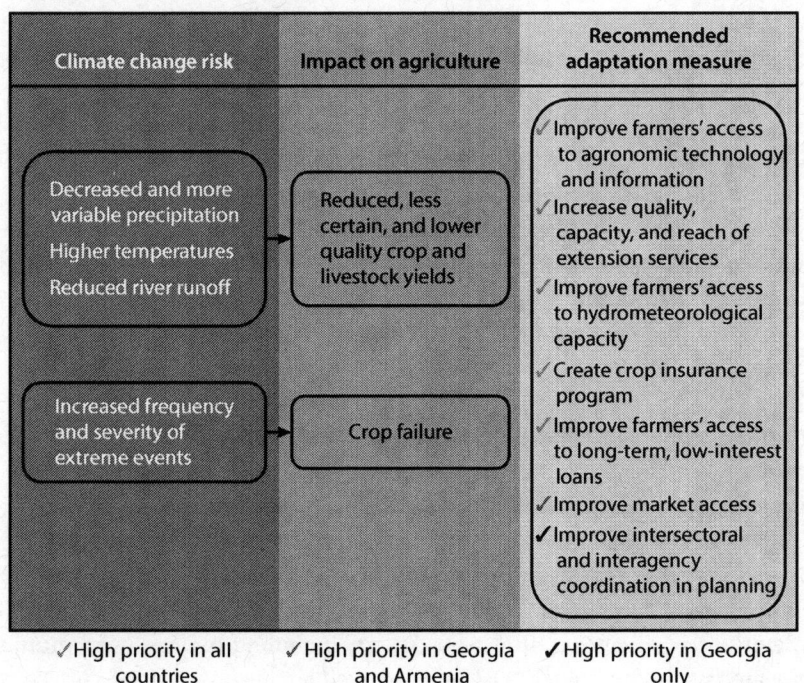

Figure O.3 Priority Measures in Agricultural Regions

High priority in all regions	High priority in irrigated regions	High priority in mountainous regions
1. Optimize agronomic practices, including fertilizer application 2. Improve crop varieties, particularly drought-tolerant crops	1. Optimize application of irrigation water 2. Improve irrigation water availability, rehabilitate irrigation capacity 3. Improve irrigation techniques 4. Rehabilitate water reservoirs	1. Adjust crop variety based on elevation 2. Research and improve livestock nutrition, management, and health 3. Construct small-volume reservoirs for water storage 4. Improve drainage infrastructure 5. Reduce erosion and practice soil conservation
Other measures	1. Establish agribusiness; assist with business plans 2. Create larger-scale farms (consolidate) 3. Establish reforestation 4. Create windbreaks	

measures recommended for Armenia, Azerbaijan, and Georgia were found to be common: (1) improving farmer access to agronomic technology and information; (2) increasing the quality, capacity, and reach of extension services; (3) improving hydrometeorological capacity and farmers' access to it; (4) establishing a crop insurance program; and (5) improving farmer access to long-term, low-interest loans.

Similarly, several priority adaptation measures at the agricultural region level were also commonly identified across neighboring countries and agricultural regions, for example, the optimization of agronomic practices and improvement of crop varieties. The similarities between the adaptation measures identified at the national and agricultural region levels suggest that a collaborative, regional approach for adaptation might provide significant economies of scale, save resources, and enhance opportunities to more effectively manage shared water resources.

Coordination of Water Resources Management among South Caucasus Countries

With due respect to riparian rights and the countries' individual needs, coordination of water resources management has the potential to greatly reduce the impacts of climate change on the agriculture sector and thus to increase shared benefits. However, regional water management planning necessarily requires consideration of nonagricultural water users, including hydropower, municipal/urban water supply, and industrial users, as well as the maintenance of ecological flows and control of flooding across national boundaries. Coordinated water management has the potential to provide three key benefits:

Pursuit of transboundary integrated water resources management for hydropower development and water supply management can create new opportunities for storage and power generation. It presents an opportunity to optimize water use across all demand categories, for example, throughout the Kura-Araks basin, with particular benefits to agriculture and hydropower. Where practicable or feasible, co-managing reservoirs as part of an integrated river basin system could provide multifaceted opportunities to all beneficiaries.

Regional water quality management and monitoring can provide economic and environmental benefits across the basin. Interventions, such as improved on-farm drainage combined with fertilizer and pesticide management by upstream beneficiaries, can have considerable downstream benefits, such as increased production at reduced cost and improved water quality. Protection of riverine aquatic ecosystems, and the Black and Caspian Seas, will require collaboration, the payoff to include better water quality for all uses.

Regionally executed climate change adaptation measures in the water sector stand to provide multiple benefits. Climate change may exacerbate regional competition over the use of available water. It is recommended that national adaptation strategies not ignore neighboring country strategies that may risk ineffective outcomes over the use of these resources and are needed to maximize shared benefits of adaptation at the regional level. For example, increasing irrigation efficiency by an upstream riparian will provide both economic benefits to local communities and greater water availability downstream.

Collaboration on Agricultural Research and Extension
Similar climate, land, ecology, and crop patterns suggest that similar varieties are likely to be well-suited across the three subject countries, providing opportunities to share costs and benefits of research and outreach, as conditions permit. Climate change will alter crop suitability conditions, suggesting that countries could work together to conduct research on new varieties that are adapted for the forecasted climate changes of the region, which likely will be hotter overall, drier in the lowlands, and wetter in the higher altitude areas. The Consultative Group on International Agricultural Research (CGIAR) system could be a technical partner that could catalyze an integration of respective national agricultural research programs. In addition and as desired, countries could agree to undertake agronomic, water-saving technologies, plant protection, and crop variety research on a single crop or family of crops (for example, cereals, fruit trees, vegetables). Results could be shared through national or regional information dissemination. Any level of coordination of research would reduce risks of duplication and optimize the use of limited research budgets.

Provision of climate services to farmers, including enhanced weather forecasting, could be usefully pursued at the regional level. Distribution of accurate and timely local weather information to farmers is an example of a climate adaptation service that could be effectively coordinated across any of the three countries, particularly to ensure optimal weather forecast sharing in cross-border and adjacent agricultural regions.

Regional Cooperation for Strategic Planning

Pursuing these plans through regional collaboration will require three types of capital: human, financial, and information. Human capital in the region resides in domestic institutions, international finance institutions, nongovernmental organizations (NGOs) such as the World Wildlife Fund (WWF), and partnerships established through cooperative efforts as embodied in this study. All three countries have a strong tradition of research and extension but they are not yet oriented toward adapting to current and forecasted climate challenges. Currently, this human capital is not coordinated across countries, which has led to unnecessary duplication and as a result the commonalities in the region of climate, agro-ecosystem types, and shared water resources have not been exploited efficiently and effectively.

Experience shows that critical steps in preparing an integrated strategic plan need to be based on sound analysis and a deep understanding of the challenges, opportunities, and potential tradeoffs of such a plan. The result envisaged would be strategic programming for investments that capture green, clean, resilient, and inclusive growth options. Such planning processes require improved awareness among the various stakeholders (for example, senior government officials in key line ministries, civil society, parliamentarians, and the private sector including the farming community) on the need for changed development pathways. Analytical work articulating the potential costs of current climate risks to development goals and the costs and opportunities to move to greener, climate-smart options are also important for elevating these development options into decisions at key ministries. Important in overall planning of these investments are access to quality information and analysis and a systematic climate risk assessment using historical and projected changes in climate, their impacts, and options for minimizing risks to development. These steps provide important signals to donors of the readiness of recipient countries to access funds that finance and support adaptation measures.

Financial capital flows from local and bilateral and multilateral sources (such as the U.S. Agency for International Development, European Union, World Bank, International Fund for Agricultural Development, and the United Nations Development Programme, among others) could be a critical component of the plan. The current efforts of these organizations so far appear uncoordinated, where their comparative advantages could be better harmonized and exploited in support of more consistent and continuous efforts toward climate adaptation. In addition, the term "climate finance" describes the financial flows that incentivize and enable projects and programs aimed at enhancing climate change mitigation, adaptation, policy, and capacity-building. New opportunities in climate finance may provide a strong incentive to recognize the benefits of regional economies of scale, with an initial focus on national governments. Chapter 7 provides additional information on development of climate finance proposals.

Information resource needs in the South Caucasus region include natural resource quality and quantity information, economic information on the efficiency of adaptation measures within and across sectors, and options for incorporating internationally available adaptation measures (such as new crop varieties). The fulfillment of information needs depends on three sources: (1) analytic work (from this study and the other ongoing studies detailed in chapter 7), (2) qualitative information (from farmers and local experts and policy makers), and (3) globally available information ranging from pure data to academic knowledge. The last category in particular is an excellent starting point for region-scale collaboration to learn about such options as climate-smart agriculture including, for example, conservation tillage.

Next Steps: Developing an Action Plan

The urgent need for national and subnational actions cannot be overemphasized. The next steps for the national programs are to incorporate the priority measures indicated earlier. More detailed information is available for each country, but it must be coordinated and analyzed. Taking no action is not an option because of the cost. The countries of the South Caucasus should address climate change through collaboration on issues such as climate-related data sharing and crisis responses. Furthermore, the management of shared water resources would be an economically efficient response to climate change that would address food security.

Table O.2 can serve as a starting point for pursuing the three strategic elements of a regional plan for the greater South Caucasus to adapt the agriculture

Table O.2 Summary of a Regional Agriculture Sector Climate Adaptation and Mitigation Plan for the Three South Caucasus Countries

Strategic elements	Objectives	Potential issues and barriers to overcome	Responsible authority	Existing models for collaborative efforts	Key outputs
1. Coordinated management of water resources among South Caucasus countries	• Reduce impacts of climate change to agriculture sector • Increase shared benefits, particularly for storage/ hydropower development • Maintain ecological flows and water quality	• Riparian rights • National–level needs may conflict • Water flow and quality data may be inconsistent	• Initially national ministries • Once established, a joint or several Basin Authority(ies)	• Interstate Commission for Water Coordination (Central Asia) • International Commission for the Protection of the Danube River	• Co-managed Basin Authority(ies) for each basin established • Collaborative management capacity developed • Knowledge and decision-support products disseminated and maintained • Kura-Araks River Basin Management Plan

table continues next page

Building Resilience to Climate Change in South Caucasus Agriculture
http://dx.doi.org/10.1596/978-1-4648-0214-0

Table O.2 Summary of a Regional Agriculture Sector Climate Adaptation and Mitigation Plan for the Three South Caucasus Countries *(continued)*

Strategic elements	Objectives	Potential issues and barriers to overcome	Responsible authority	Existing models for collaborative efforts	Key outputs
2. Collaboration on agricultural research and extension	• Reduce impacts of climate change on agriculture sector • Jointly access and influence CGIAR research • Gain economies of scale in extension • Enhance access to current technologies	• Rights to agricultural technologies • Current extension may be poorly subscribed or relied upon by farmers	• National Ministries of Agriculture and Education	• CGIAR components	• Research results shared across the three countries • Country-level extension programs incorporate the new research results in demonstration plots and trainings • National-level research better coordinated
3. Enhanced climate services (including short-term forecasting and climate projections) provision to farmers pursued at the regional level	• Reduce impacts of climate change to agriculture sector • Expand capabilities of hydromet services for the region	• Existing monitoring equipment and data collection may be inconsistent • Intellectual rights to data can be complicated and can limit sharing of products across countries and with farmers	• National hydromet institute, in partnership with users and stakeholders at various scales	• Climate Services Partnership • Caribbean Institute for Meteorology and Hydrology • International Research Institute for Climate and Society (e.g., see http://scalingup .iri.columbia .edu/index.html) • AGRHYMET Regional Center (extreme events forecasting)	• Distribution of accurate and timely local weather information to farmers • Creation of new long-term and extreme event forecasting capabilities for regional purposes

Note: CGIAR = Consultative Group on International Agricultural Research; hydromet = hydrometeorological.

sector to climate change. Although there are many challenges to achieving these objectives, fortunately a wide range of existing "models" of regional-scale institutional arrangements exist throughout the world, encompassing the scope of regional cooperation for water resources planning, agricultural research and extension, and enhanced hydrometeorological service development and data provision.

CHAPTER 1

Introduction and Reasons for Action

Agriculture has traditionally been an important part of the economies of the South Caucasus region. In 2011 agriculture contributed 28 percent of gross domestic product (GDP) in Armenia, 16 percent in Azerbaijan, and 22 percent in Georgia (World Bank 2013). Although the agriculture share of GDP has declined in the three countries over the past decade, all three are still agrarian societies. The·main significance of the agriculture sector is its role in employment: it has provided 40 percent or more of total employment in recent years. However, the rural populations in these countries remain poor, with rural poverty rates in 2008 of 28 percent in Armenia, 19 percent in Azerbaijan, and 28 percent in Georgia. Although more recent data are not available for all countries, rates in the region appear to be on the rise (World Bank 2013). These rural populations are therefore highly vulnerable to any climatic event that affects the agriculture sector.

Climate change is a phenomenon that could trigger a greater severity and frequency of the types of events that currently challenge agricultural production, including heat waves, floods, and droughts, as well as changes in overall temperature and precipitation regimes that affect crop and livestock productivity. At the same time, climate change can also create opportunities, particularly in agriculture. Increased temperatures can lengthen growing seasons for some crops, higher carbon dioxide concentrations may enhance plant growth, and, in some areas, rainfall and the availability of water resources can increase as a result of climate change.

Adaptation planning is challenging because of uncertainties in climatic developments and their locally specific impacts, which makes it difficult to identify the optimal changes in agricultural systems. To be successful, adaptation planning should start early and be sufficiently flexible to address these variables. Accordingly, this work sets out to identify "win-win" or "no regrets" adaptation responses that are robust under a range of different future climate scenarios and contribute to increasing resilience to present day climate challenges, such as droughts, floods, and increased heat stress. Wherever possible, this work also tries to identify "win-win-win" adaptation options that might also reduce greenhouse gas (GHG) emissions.

Overview of Geography, Climate, and Crops in Study Countries

Map 1.1 presents an overview of the geographic scope of the study, identifying the key agricultural regions within each of the three countries. Baseline agricultural conditions, climate change impacts, and adaptive options are similar within each of the regions in ways that are important for developing a specific adaptation plan. The darker areas in map 1.1 are areas of high elevation (typically characterized by mixed livestock/cereal and some high-value fruit tree production), and lighter areas are low elevation (typically characterized by irrigated high-value vegetables and fewer cereals and in the case of Azerbaijan, potential for cotton production). Contiguous transboundary areas of high elevation are common throughout the region.

In each of the three countries, the study focused on selected crops (not more than seven due to resource constraints). The crops were different in each country but in all cases selection was based on the following criteria: (1) widely grown; (2) economically important to the country; (3) potentially sensitive

Map 1.1 Agricultural Regions of the South Caucasus

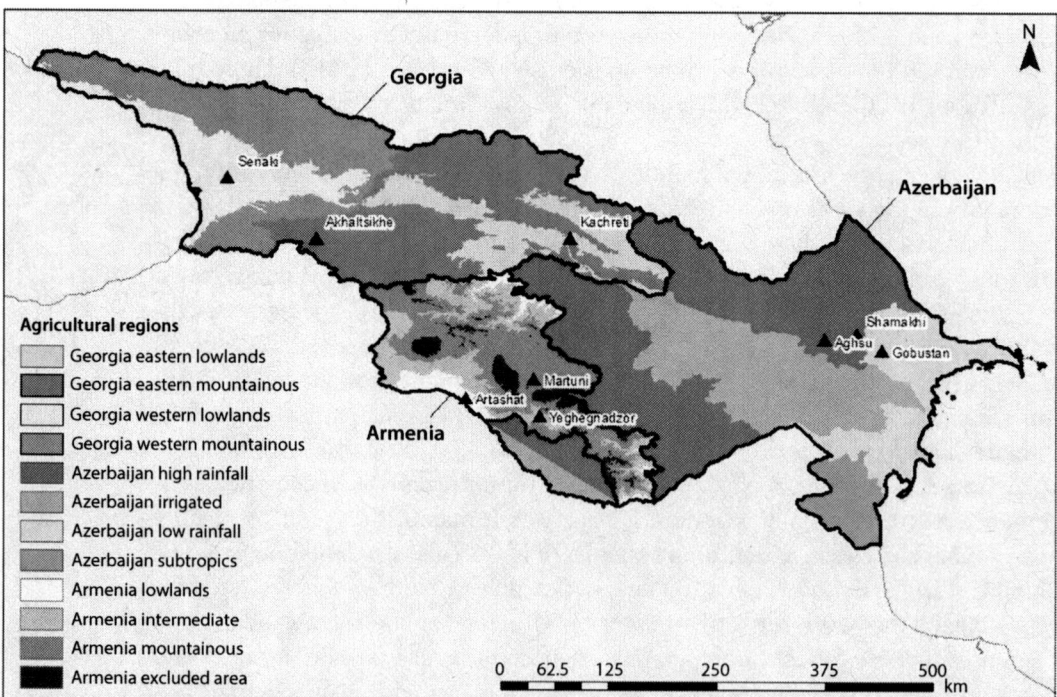

(either positively or negatively) to temperature or water stress aspects of climate change; (4) well supported by data for domestic yield, cropping patterns, and phenology; and (5) broadly reflecting a mix of primarily irrigated and primarily rainfed crops. Furthermore, to ensure a wide variety, the list included one or two representatives from each of the following groups: (1) cereals, (2) tree crops, (3) vegetables, and (4) forage crops or natural pastures. As shown in table 1.1, wheat, grape, and potato were selected as focus crops by all three countries; corn, tomato, alfalfa, and pastures were selected by two countries; and mandarin orange, apricot, watermelon, and cotton were selected in one country.

The time frame of the study is the current period, 2013 through 2050. The logic for holding the time horizon to 2050 is that virtually all measures considered by the study, including newly constructed irrigation infrastructure, would have reached the end of their useful life by 2050. Nonetheless, because recent research suggests that the potential for dramatic climate change is greater after 2050, national institutes and ministries must periodically update this analysis as the mid-century approaches and climate change unfolds.

Stakeholder consultations, particularly those with farmers, were conducted throughout the region. Map 1.1 indicates the nine locations in the region (three in each country) where farmer stakeholder workshops were conducted to discuss

Table 1.1 Crops Selected for Modeling in Each Country

Irrigated/rainfed	Crop	Armenia	Azerbaijan	Georgia
Irrigated	Alfalfa	X	X	
	Apricot	X		
	Corn		X	X
	Cotton		X	
	Grape	X	X	X
	Mandarin orange			X
	Potato	X	X	X
	Tomato	X		X
	Watermelon	X		
	Wheat	X	X	X
Rainfed	Alfalfa	X	X	
	Apricot	X		
	Corn		X	X
	Cotton		X	
	Grape	X	X	X
	Mandarin orange			X
	Pasture		X	X
	Potato	X	X	X
	Tomato	X		X
	Watermelon	X		
	Wheat	X	X	X

Building Resilience to Climate Change in South Caucasus Agriculture
http://dx.doi.org/10.1596/978-1-4648-0214-0

the challenges presented by climate change, to jointly identify viable adaptation measures, and to evaluate and prioritize these measures based on cost, feasibility, and potential of improving agricultural production in view of the challenges of climate change.

Current climate data show great variation within the three countries, owing mostly to wide variations in elevation and the effect of mountains on precipitation patterns (for example, rain shadow effects). While there is wide variation within each country, there are great similarities across the region; thus, most climate classification systems assign the three countries to a similar climate type. The Köppen-Geiger Climate Classification System (KGCCS),[1] which combines average annual and monthly temperatures and precipitation and the seasonality of precipitation in a single index, is one of these. Map 1.2 provides a summary of the KGCCS for the South Caucasus countries, for current (a) and projected (b) climate conditions, with resolution at roughly a 50 × 50 kilometer (km) grid. In map 1.2a the majority of the area is in the purple and black regions, which consists of a "snow" climate region that is "fully humid" with a "warm summer." Azerbaijan in the eastern area of the map is an exception, however, with the tan area (corresponding to lowland plains) representing an "arid steppe" region. Southwestern Armenia also has areas with this climate classification, in the highly productive Ararat Valley agricultural region. These similarities in current climate create opportunities for sharing results of agricultural research, particularly focusing on crop varieties that thrive in these climatic zones.

Map 1.2 Observed and Forecasted Köppen-Geiger Climate Classification for Azerbaijan, Georgia, and Armenia

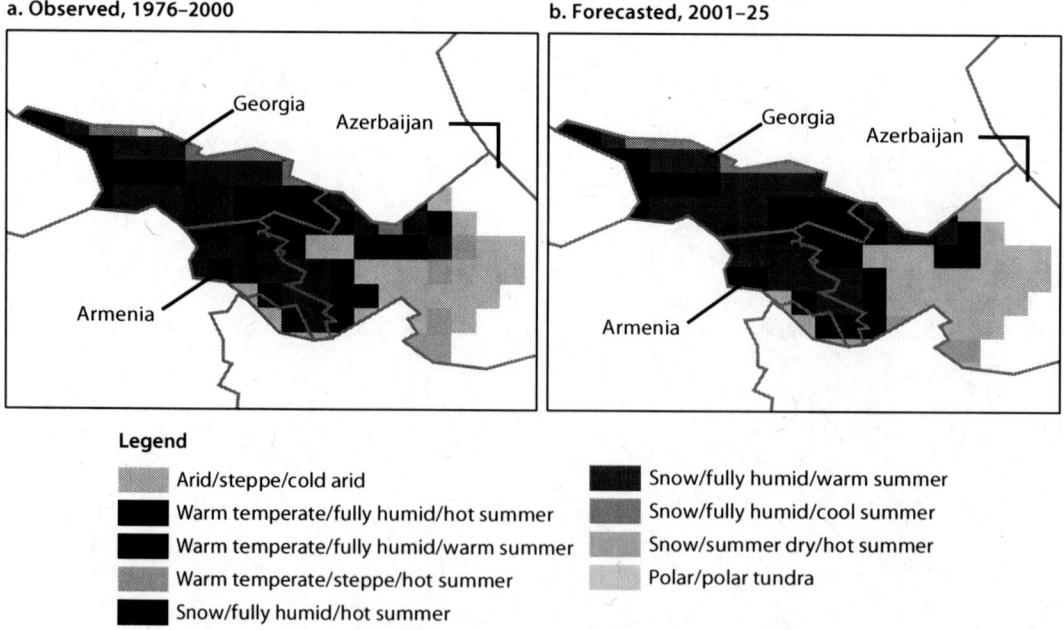

a. Observed, 1976–2000

b. Forecasted, 2001–25

Legend

Arid/steppe/cold arid	Snow/fully humid/warm summer
Warm temperate/fully humid/hot summer	Snow/fully humid/cool summer
Warm temperate/fully humid/warm summer	Snow/summer dry/hot summer
Warm temperate/steppe/hot summer	Polar/polar tundra
Snow/fully humid/hot summer	

Sources: Author mapping of data described in Rubel and Kottek 2010; data provided at http://koeppen-geiger.vu-wien.ac.at/shifts.htm.

The implication of climate change, even when looking ahead only about 10 years to 2025 (map 1.2b), is for a dramatic expansion of the arid (tan) regime in Azerbaijan, particularly in the central area of the country, as well as changes in the temperature regime throughout Armenia and Georgia. The warm temperate and snow regimes largely cross national boundaries throughout the region, and these transitional areas with similar climate could be identified as good candidates for adaptation measures, for example, for variety adaptation trials.

These similarities in climate are also expressed in maps of landscape and ecology type. Map 1.3 shows the high elevation thermo-moderate and humid mountain landscape (classification O, in orange) is common to all three countries, including the northern and southern bands in Georgia; northern Armenia near the Georgia/Azerbaijan border; and the north, west, and extreme south portions of Azerbaijan. There is a high prevalence of the lower elevation subtropical and plain and hilly landscape (classification E, shown in light blue) in central and southern Azerbaijan, including the Absheron Peninsula, which corresponds well to the intensely irrigated areas; such landscape is not found in the other two countries. The high elevation temperate semi-arid mountain

Map 1.3 Landscape Types in Armenia, Azerbaijan, and Georgia

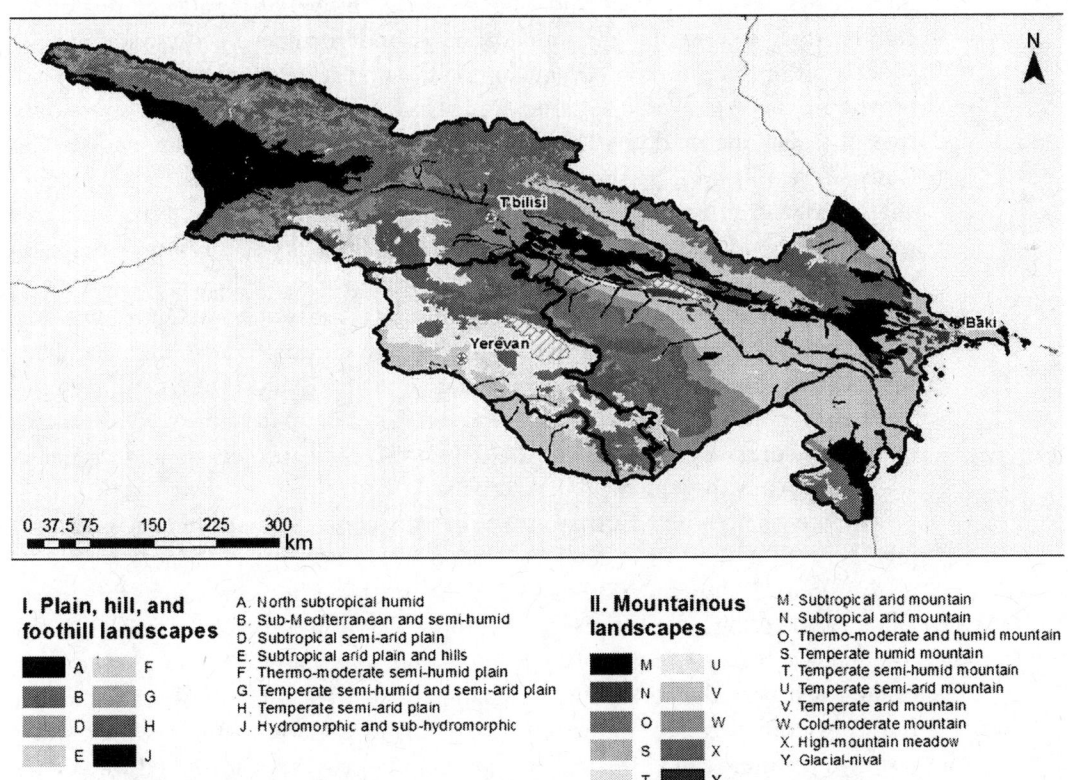

I. Plain, hill, and foothill landscapes

A. North subtropical humid
B. Sub-Mediterranean and semi-humid
D. Subtropical semi-arid plain
E. Subtropical arid plain and hills
F. Thermo-moderate semi-humid plain
G. Temperate semi-humid and semi-arid plain
H. Temperate semi-arid plain
J. Hydromorphic and sub-hydromorphic

A F
B G
D H
E J

II. Mountainous landscapes

M. Subtropical arid mountain
N. Subtropical arid mountain
O. Thermo-moderate and humid mountain
S. Temperate humid mountain
T. Temperate semi-humid mountain
U. Temperate semi-arid mountain
V. Temperate arid mountain
W. Cold-moderate mountain
X. High-mountain meadow
Y. Glacial-nival

M U
N V
O W
S X
T Y

Sources: ©Industrial Economics. Used with permission; reuse allowed via Creative Commons Attribution 3.0 Unported license (CC BY 3.0). Country boundaries are from ESRI and used via CC BY 3.0.
Note: km = kilometers.

landscape dominates in Armenia and has some prevalence in southern Georgia near the Armenian border, as well as small patches in Azerbaijan, in the north-eastern portion of Nakhchivan, the far north central areas near the Russian border on the north slope of the Greater Caucasus, and the far southeast near the Iranian border.

The detailed maps provide an excellent basis for identifying areas with similar ecology (natural vegetation, climate, and soil characteristics) which parallel characteristics for crop suitability. These delineations were used as a guide in the study for agriculture sector climate change adaptation. As indicated in map 1.2 (climatic regions), these areas can also be used to identify "transnational regions" for testing and demonstrating promising varieties and also for developing site-specific agronomic practices to enhance productivity and profitability.

The eco-region maps also provide insights concerning the opportunities for mitigating GHG emissions. A key factor in establishing the GHG mitigation potential of land is the ability of the soil to store carbon—for example, a healthy pasture is better equipped to sequester carbon, first in plant material and ultimately in storage of organic carbon in soils. With some exceptions, warmer and wetter areas are more ecologically productive, making them better able to capture and store atmospheric carbon. Therefore the maps provide a starting point for the design of national- and regional-scale mitigation strategies related to carbon storage, particularly in unmanaged or undermanaged ecosystems.

In addition to common climate and landscape regimes, a significant characteristic of the region is its shared water resources. Two major transboundary river basins—the Mtkvari/Kura and the Araks/Aras, both flowing to the Caspian Sea, as well as several smaller sub-basins within this larger basin—characterize the South Caucasus area. The transboundary nature of water resources provides an opportunity to examine regional water resource management among the riparians.

A series of smaller sub-basins are also important areas of shared transboundary water resources, in part because they are in high elevation areas that may have significant potential for reservoir construction, where such storage could be used for irrigation and hydropower development. These include the Alazani/Ganikh basin in Georgia and Azerbaijan, the Khrami-Debed in Georgia and Armenia, and the Aghstev in Armenia and Azerbaijan.

Climate change will further stress water resources in the South Caucasus region. Precipitation is projected to decline and temperatures to increase, resulting in runoff decline by 2050 or sooner. At the same time, crop water demands will increase due to the higher temperatures. The transboundary nature of water resources is coupled with the high likelihood that water flow and volumes in general will be reduced by climate change presenting the risk of conflicts over the ever-more precious water resources. However, coordinated regional water resource planning can alleviate these conflicts. For example, increased water storage is almost always more efficiently constructed in higher elevation areas, where natural terrain can be exploited to create reservoirs and where the steeper terrain creates greater potential for hydropower generation at the reservoir outlet.

An added benefit is that higher elevations are cooler and evaporative loss from reservoirs is reduced. A well-structured multinational water management system holds promise for the higher elevation countries (mainly Armenia but also parts of Georgia) to develop these storage opportunities and sell both water and hydropower throughout the region, in exchange for other trade considerations. Managing the reservoirs as part of an integrated river basin system would provide benefits for all riparians.

Characteristics of the Agriculture Sector

The typical agricultural system in these countries is subsistence or semisubsistence mixed crop production integrated with small-scale livestock production. In fact, livestock production has long been an important component of the agricultural economies across the region. It is often dependent on communal grazing lands that are usually degraded in terms of both land and vegetation due to overgrazing and consequent soil erosion. Pastures dominate in the high-altitude regions, particularly along the southern face of the Greater Caucasus, which runs through northern Georgia and Azerbaijan, while in the lower altitude areas mixed farming dominates and is particularly prevalent in rainfed areas of Azerbaijan and Armenia.

The three countries also rely heavily on irrigation for high-value crop production, where, according to the Food and Agricultural Organization of the United Nations (FAO) AquaStat, the agriculture sector is by far the largest consumptive user of water (FAO 2013). Map 1.4 illustrates the current reliance on irrigation for all three countries. The irrigated lands are more extensive in Azerbaijan than in the other two countries, but this observation ignores important aspects of crop patterns. In Armenia about 80 percent of the overall value of crop production occurs on irrigated lands. In Georgia much of the high-value agricultural production (for example, grape) occurs in areas that are currently classified as semi-arid, but where precipitation and runoff are both expected to decline as a result of climate change. In addition, although many areas of Georgia are currently equipped for irrigation, on-farm water delivery still suffers mostly due to the need for rehabilitation of infrastructure.

Agriculture in the region is predominantly carried out by rural households where some land has been distributed from former state-run farms and collectives after the Soviet breakup. These smallholder farmers usually have fragmented land holdings of 1–3 hectare (ha) in several plots, thus facing constraints of small areas, limited profits, and scarce financial means. Having been former employees of the state farms where they were delegated with specific and often nonagricultural tasks, many farmers lack farming backgrounds. They need tailored advice; however, no effective and efficient extension system is in place to provide the service on required scale and quality.

Similarities in agricultural land characteristics and the prevalence of irrigated and mixed livestock/crop production patterns across the region present important opportunities for collaboration. In particular, as all three countries seek to address development and adaptation deficits in their agricultural land management

Map 1.4 Irrigated Lands in the South Caucasus

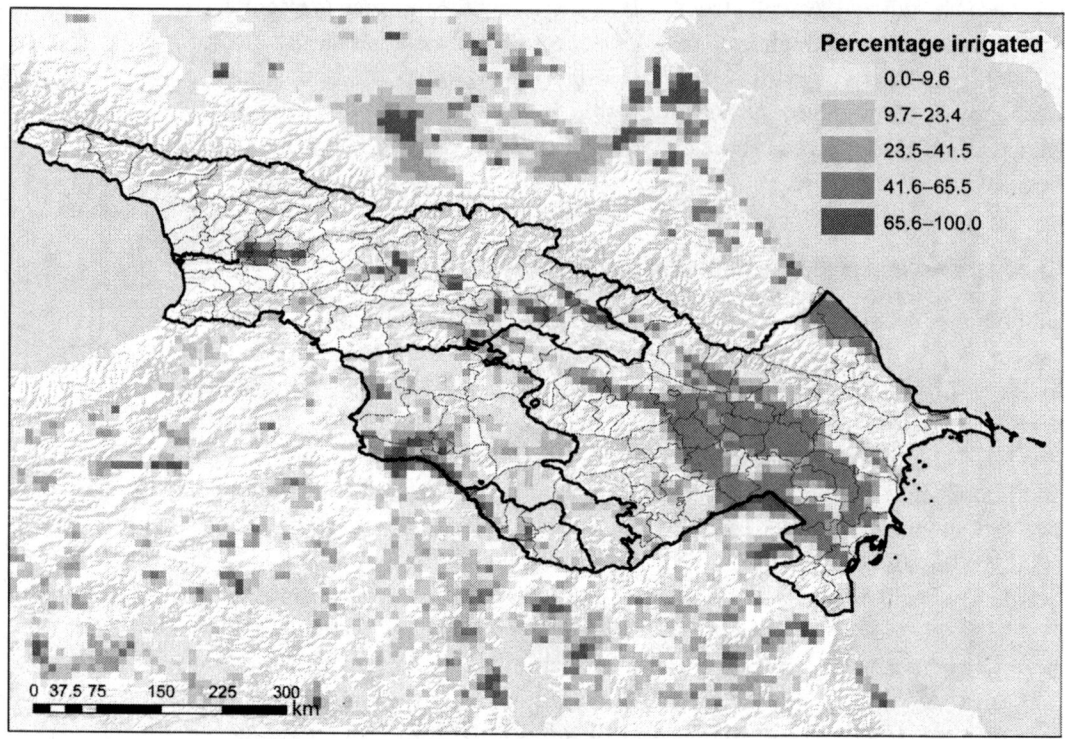

Sources: ©Industrial Economics. Used with permission; reuse allowed via Creative Commons Attribution 3.0 Unported license (CC BY 3.0). Country boundaries are from ESRI and used via CC BY 3.0.
Note: km = kilometers.

practices—for example, by learning and applying modern agronomic practices and improving water-use efficiency—a cooperative approach can yield cost savings and significant transboundary spillover benefits compared to each nation pursuing these agriculture sector improvement measures independently.

In addition, the governments of Armenia, Azerbaijan, and Georgia have recently developed agricultural policy documents that outline some of the primary challenges facing the sectors. These documents identify the following key issues, suggesting that the countries of the South Caucasus face many of the same challenges (World Bank 2007; Azerbaijan Republic 2008; Urutyan and Thalmann 2011; FAO 2012; Georgia Ministry of Agriculture 2012; IFAD 2012):

- Lack of effective extension and research services
- Insufficient and inadequate use of fertilizers and pesticides
- Lack of mechanization and/or outdated agricultural machinery and equipment
- Inadequate irrigation coverage or inefficient irrigation practices
- Soil erosion, land degradation, salinization, and/or limited land resources
- Natural disasters
- Limited water resources
- Lack of market access

Agriculture Sector Capacity to Adapt to Climate Change

A country's capacity to adapt to climate change reflects a wide range of socioeconomic, policy, and institutional factors, at the farm, national, and regional levels. Considerations in determining the variation in adaptive capacity across a country or a region also include the current climate, social structures, institutional capacity, knowledge and education, and access to functioning infrastructure. Specifically, marginal areas under rainfed production will have less adaptive capacity than areas that are irrigated and more productive. In addition, financial resources are key in determining adaptive capacity, as most planned adaptations require investment. Currently, the region's countries rank low in their agriculture sectors by many factors that determine a country's overall adaptive capacity.

In any country the level of adaptive capacity in the agriculture sector is characterized by a number of factors: (1) high level of functionality in the provision of hydrometeorological and relevant geospatial data to farmers to support good farm-level decision making, (2) provision of other agronomic information through trained extension agents and effective extension networks, (3) in-country research oriented toward innovations in agricultural practices in response to forecasted climate changes, (4) well-maintained and managed water collection and distribution infrastructure that meets the needs of the farming community, and (5) systems in place to resolve conflicts between farmers and other users over water allocation.

Some of these conditions exist in the study countries, but most are inadequate or lacking in the following ways: (1) inefficient and ineffective extension service, (2) weak agricultural research-extension linkage, (3) limited access to rural finance, (4) limited crop insurance, (5) poor access to meteorological data, and (6) poor market access. These conditions are described as follows.

The current agricultural extension service is not oriented toward ameliorating risks from climate. While many farmers are aware of the extension service, only a few have access to these services or can make use of them. Furthermore, the current extension service has limited capacity to advise on adapting agricultural systems to the climate risks outlined in this study. Farmers in the region indicate that they would benefit highly from a well-functioning, effective extension service. In agriculture, climatically (as in weather) induced risks are inherent to the system. Farmers may be risk-averse but they need knowledge and skills to manage their risks.

Agricultural research-extension linkages, if not lacking, are weak and erratic. Agricultural research institutes remain an important part of the agricultural bureaucracy in these former Soviet countries, but these institutions have not yet given priority to and focused on climate change as a major risk to agricultural production, and their research is not coordinated with the extension service as it should be. Further, research could be better focused on leveraging advances in crop varieties and farming practices proven to be effective in other countries, as well as coordinating with the extension service to carry on-farm adaptation trials and then demonstrate these results locally.

Building Resilience to Climate Change in South Caucasus Agriculture
http://dx.doi.org/10.1596/978-1-4648-0214-0

Farmers' access to rural credit is limited. Farmers note difficulty in obtaining long-term, low-interest bank loans for agriculture. These financial constraints limit mechanization of production on most small farms. While government-sponsored credit subsidy programs exist or are being planned, farmers consistently emphasized that even if they want to invest in equipment and agricultural inputs to improve their practices, financial issues are the major bottleneck. Many of the credit issues in the region are also linked to weaknesses in land policy and land markets.

Crop insurance is either not affordable or not available. Both hail and spring frost are major issues for farmers in the region, with estimates of annual losses on the order of 10 percent of annual production for some crops, which may account for as much as US$100–150 million in annual losses in Armenia alone.[2] Even where insurance is available, farmers are generally unable to afford it. Subsidized disaster relief programs, including insurance, would greatly stabilize their incomes and improve their capacity to re-invest in farming.

The ability to collect, generate, and disseminate meteorological data to farmers is either inadequate or lacking. Current capacity in hydrometeorological institutions needs improvement, as farmers lack basic climatic and meteorological data for their regions—except weather forecasts on public TV—that they can utilize in operational farm management. Specifically, most farmers do not have the financial means to buy specific hydrometeorological services or related equipment.

Agricultural marketing is a common problem. More must be done to improve markets if the agriculture sector potential is to be realized. Several projects that targeted marketing were financed by international donors, but the problem still prevails in the region where a large portion of farmers practice subsistence and semi-subsistence farming with poor market links and outdated varieties. The farming community as a whole complains about the following interlinked problems, some of which extend beyond but are related to marketing: (1) low commodity prices, (2) inability to market the produce even though the market is not saturated, (3) distance to the markets, and (4) lack of access to agro-processing. The underlying reasons include poor quality of the products due to poor production and post-harvest practices, timing of marketing, lack of storage facilities, lack of adequate information related to production and marketing, and problems regarding transportation.

An Approach for Adapting to Climate Change in the South Caucasus

The key insights from the study related to land, water, climate, ecological conditions, and development and adaptation capacity throughout the South Caucasus region are as follows:

• ***Land.*** Crop suitability of land resources, as evaluated by FAO and the World Bank, is quite similar across the three countries: Irrigation is important for high-value crop production in the region, with Azerbaijan having the largest irrigated area.

- **Water.** The key Kura-Araks river system is transboundary, so issues of management of water resources, both quantity and quality, demand a supranational approach. Reflecting this insight, in this study the transboundary river system was modeled as a multicountry unit, encompassing water demand and supply in the agriculture, urban/municipal, and hydropower sectors.

- **Climate.** Climate is diverse within the countries due to undulated topography and large variability in elevation. However, the climatic regions in these three countries are very similar, as is the projection of climate change.

- **Ecology.** Ecological regions are transboundary in nature, with a high degree of common landscape types, particularly across the mountainous regions. All three countries exhibit similar ecotypes with elevated areas of low and high rainfall, including both steeply sloping areas and elevated meadows and plains.

- **Adaptive capacity.** As noted, all three countries have common development and adaptation deficits in the agriculture sector. The development deficit refers to a low level of adoption and knowledge of modern agricultural practices, as well as prevailing financial and market constraints. The *adaptation deficit* refers to inadequate adjustment to current climate conditions, including high vulnerability to extreme weather events, poor access to weather and climate information, and poor uptake of new technologies and information that can ameliorate impacts of climate. Addressing *both* is critical to improving the resilience of the region's agriculture to climate shifts.

The key insight of this book is that the regional nature of the natural resources and current adaptive capacity of the countries makes a multicountry transboundary approach to management advantageous to neighboring countries. There is great potential for cost savings and significant spillover benefits to accrue to each country if measures addressing these deficits are pursued as part of a collective program with direct reference to their shared geographies and natural resources, rather than in isolation within each country's national programs. At present, natural resources are managed with a rather narrow, national perspective. The impact of climate change in the region is likely to make suitable land and high-quality water resources more scarce for all, putting new pressures on ecological resources and presenting more risks than opportunities. Neighboring countries stand to gain in the agriculture and water resources sectors from cooperation, much more so than if national interests are pursued without regard for transboundary implications.

Notes

1. The effort is described and applied to the globe in Rubel and Kottek (2010). The Köppen-Geiger system classifies land areas based primarily on their climate characteristics, and the system is based on the concept that native vegetation is the best

expression of climate. Therefore climate zone boundaries are made with vegetation distribution in mind. The results can be generated for both historic and projected climate conditions, and both historic and projected results are readily available for the globe via a website, http://koeppen-geiger.vu-wien.ac.at/shifts.htm. The key advantages of the Köppen-Geiger system are that it is well known and often cited within the climate change and other literatures and that it is widely available and readily replicable.

2. Estimates of annual losses are from World Wildlife Fund Norway (WWF 2009) and from discussion during the first farmer consultation of the study with independent consultant Tigran Kalantaryan, who facilitated the farmer consultations.

References

Azerbaijan Republic. 2008. "The State Programme on Reliable Food Supply of Population in the Azerbaijan Republic (2008–2015)." Government of Azerbaijan, Baku (accessed December 9, 2013), http://tinyurl.com/l3mcpz7.

FAO (Food and Agricultural Organization of the United Nations). 2012. "FAO Republic of Armenia Country Programme Framework 2012–2015." FAO, Rome (accessed December 9, 2013), http://www.fao.org/fileadmin/user_upload/FAO-countries /Armenia/Armenia_CPF_FINAL_English.pdf.

———. 2013. "AQUASTAT." FAO, Rome (accessed December 9, 2013), http://www.fao .org/nr/water/aquastat/main/index.stm.

Georgia Ministry of Agriculture. 2012. "Agricultural Development Strategy 2012–2022." Government of Georgia, Tbilisi (accessed December 9, 2013), http://www.fao.org /fileadmin/user_upload/eufao-fsi4dm/doc-training/Agricultural_development _strategy.pdf.

IFAD (International Fund for Agricultural Development). 2012. "Georgia Agricultural Support Project: Supplementary Financing Design Report." IFAD, Rome (accessed December 9, 2013), http://www.ifad.org/operations/projects/design/107/georgia.pdf.

Rubel, F., and M. Kottek. 2010. "Observed And Projected Climate Shifts 1901–2100 Depicted by World Maps of the Köppen-Geiger Climate Classification." *Meteorologische Zeitschrift* 19 (2): 135–41.

Urutyan, V., and C. Thalmann. 2011. "Assessing Sustainability at Farm Level Using RISE Tool: Results from Armenia." Working Paper, International Congress of European Association of Agricultural Economists, August 30–September 2, Zurich. (accessed December 9, 2013), http://ageconsearch.umn.edu/bitstream/114820/1/Urutyan _Vardan%20_473.pdf.

World Bank. 2007. *Integrating Environment into Agriculture and Forestry Progress and Prospects in Eastern Europe and Central Asia, Volume II: Armenia—Country Review.* Washington, DC: World Bank (accessed December 9, 2013), http://www.worldbank .org/eca/pubs/envint/Volume%20II/English/Review%20ARM-final.pdf.

———. 2013. The World Bank DataBank: World Development Indicators (database). World Bank, Washington, DC (accessed December 9, 2013), http://databank.worldbank .org/data/views/variableSelection/selectvariables.aspx?source=world -development-indicators.

WWF (World Wildlife Fund Norway and WWF Caucasus Programme). 2009. "Climate Change in Southern Caucasus: Impacts on Nature, People and Society." Report, WWF Norway, Oslo. http://assets.wwf.no/downloads/climate_changes_caucasus___wwf _2008___final_april_2009.pdf.

CHAPTER 2

Framework and Program Design

The following are the main objectives of this study, which is entitled, "Regional Program on Reducing Vulnerability to Climate Change in Southern Caucasus Agriculture Systems":

- Increase stakeholder awareness of the threat of climate change to the agriculture sector and seek their input to shape the results and, in some cases, the methods applied in the study.
- Analyze the vulnerability and potential impacts of climate change on agricultural systems at multiple levels—agricultural region, national, and regional.
- Integrate agricultural sector analysis with in-depth modeling of water supply and demand, with the understanding that climate change will affect the agriculture sector both directly and indirectly through its effect on the availability of irrigation water.
- Combine biophysical modeling with economic benefit-cost (B-C) analysis and qualitative assessment to develop a prioritized menu of potential adaptation options for each subnational agricultural region and at the national level.
- Create mechanisms for fostering regional cooperation to address the potential impacts of climate change on agriculture.

Study Approach

The study was conducted in the three countries of the South Caucasus region—Armenia, Azerbaijan, and Georgia—at the country and agricultural region levels. The study scope included the agriculture, livestock, and water resources sectors, or, more succinctly, systems within the managed agriculture sector. Time and resource constraints meant that the study excluded the forestry, fisheries, biodiversity, and urban/peri-urban agricultural systems.

The study was conducted in three stages from January 2012 through April 2013. The study team followed four main steps, summarized in figure 2.1, described as follows:

Step 1: Awareness Raising and Consultation with Countries

Country notes. A "Country Note on Climate Change and Agriculture" (World Bank 2012a, 2012b, 2012c) was developed for each country as a background document for all stakeholders and to serve as an engagement tool for awareness raising and consultation. The country note provided a summary of available country-specific information with a focus on climate and crop projections, adaptation options, policy development, and institutional involvement in agriculture and climate change.

Awareness raising and consultation workshop. An initial country-level awareness raising and consultation workshop was held in each country in consultation with key stakeholders at the technical level, including local experts from national, private-sector, and nongovernmental institutions, as well as representatives of other development organizations. The objectives of the workshops were to raise stakeholder awareness of agriculture and climate change issues, discuss the country note, identify any other relevant analytical work in the country, elicit ideas on potential adaptation responses, agree on information gaps and needs for additional analysis, and identify local partners to engage in the development and implementation of country specific analytical approaches

Figure 2.1 Flow Chart of Major Study Steps

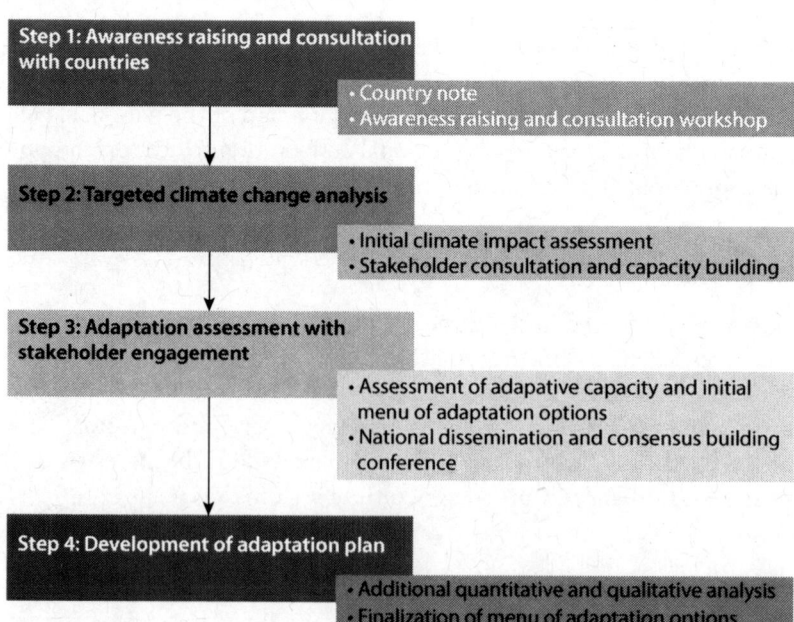

for climate change impact assessment and analysis of adaptation options, including data collection, analysis, and dissemination follow-up activities. After the awareness raising and consultation workshop was completed, an inception report was developed, which served as a work plan for the subsequent steps.

Step 2: Targeted Climate Change Analysis

Initial climate impact assessment. The study provides forecasted changes in temperature and precipitation at the agricultural region level, which are used to inform quantitative analysis of crop yield and irrigation water resource system impacts that could occur without adaptation measures.

Stakeholder consultation and capacity building. The study team solicited feedback from stakeholders on the results of the initial climate impact assessment and provided a capacity-building opportunity for local stakeholders to learn about the potential impacts of climate change on the agriculture sector.

Step 3: Adaptation Assessment with Stakeholder Engagement

Assessment of adaptive capacity and initial menu of adaptation options: After assessing each country's existing adaptive capacity, the study team developed an initial menu of adaptation options, tailored to the agricultural regions in each country. This draft menu of options was then vetted with farmers at a second stakeholder consultation.

National dissemination and consensus-building conferences. A national conference was organized in each country in order to discuss and raise awareness of the results of the impact assessment and the initial menu of adaptation options. Stakeholders worked together at the conferences to build consensus on the priorities for action and explore ways to integrate adaptation recommendations into country policies, programs, and investments. The conferences were co-hosted by the ministries of agriculture and environment along with World Bank country offices. The organizers sought high-level representation from agencies with national policy-making responsibility, such as ministries of finance and economics. Representatives of farmers and other civil society organizations also participated, and development partners who could help support adaptation actions were invited.

Step 4: Development of Adaptation Plan

Additional quantitative and qualitative analysis. The study team undertook additional analysis based on feedback received during the stakeholder consultations and national conferences. The analytic tools employed in these analyses are detailed in the following section.

Finalization of analysis and menu of adaptation options. Finally, country-specific menus of adaptation options were finalized and disseminated. The recommended options are based on the results of quantitative analysis (B-C assessments), as well as qualitative analysis by stakeholders and experts.

Analytic Tools

Overall four analytic steps, involving four types of analytic tools, were required to develop the menu of adaptation options. As shown in figure 2.2, these analytic steps were carried out sequentially from top to bottom, with the exception of the interaction between the crop and water balance modeling, which is discussed here. The tools in figure 2.2 were needed to complete the initial impact assessment in four steps as follows: (1) gather baseline data and identify major agricultural growing regions in each country, (2) develop climate projections, and (3) use baseline and climate projection data to conduct the impact assessment. The study team then carried out an additional analytic step to develop a tailored adaptation menu, specifically: (4) evaluation of adaptation options for each agricultural region in each country.

Achieving the goals stated at the beginning of this chapter dictated certain aspects of the modeling approach. For example, the study team immediately identified that a simulation modeling approach to the quantitative work would be most appropriate. Simulation modeling can be demanding—simulating the processes of crop growth and water resource availability requires extensive data inputs and careful calibration. In addition, simulation modeling can present difficult issues in modeling a future economic baseline that incorporates innovation over time in those situations where it may be important to the analysis to do so.[1]

Figure 2.2 Flow Chart of Analytic Tools for Key Analytic Steps

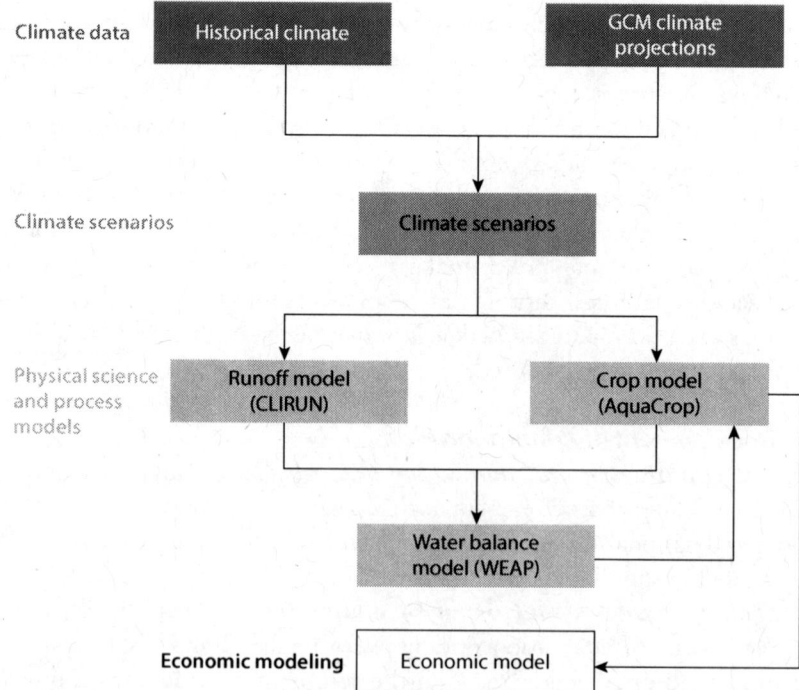

Note: CLIRUN = Climate and Runoff Hydrologic Model; GCM = General Circulation Model; WEAP = Water Evaluation and Planning System.

The payoff is that the modeling system can estimate the incremental change in crop output and water supply in response to changes in climatic conditions and agricultural and water resource management techniques. Other approaches, such as econometric and statistical models of crop yield, are often unable to incorporate adaptation or, if they do incorporate it, cannot estimate the incremental effects of specific measures.[2] A further advantage of the simulation approach is that it provides an opportunity for stakeholder involvement at several stages of the analytic process: designing scope, adjusting parameters, selecting inputs, calibrating results, and incorporating adaptation measures of specific local interest (for example, in half of the countries, hail nets, crop insurance, water storage, and improved drainage capacity were major issues, in each case involving a different pair of countries).

Analytic Step 1: Gather Baseline Data and Identify Agricultural Regions

The first analytic step involved gathering baseline meteorological, soil, and water resources data from in-country and global sources. Data requirements include the following:

- *Meteorological.* The crop modeling methodology required at least 10 years of daily historical data in the major agricultural regions of each country.
- *Soil characteristics.* Crop modeling requires data on soil type, suitability, erosion potential, and hydrology characteristics.
- *Water resources.* Water resources modeling requires at least 10 years of average daily or monthly (daily preferred) historical river flow data for gauging stations along the main stem rivers of each major drainage basin. These data were provided by in-country sources. In addition, the study obtained locations and active storage volumes of each major reservoir from in-country sources.

The station-level meteorology data provided by local sources varied in quality and comprehensiveness. While some countries had excellent data and shared the data readily with the project team, institutional capacities prevented others from providing useful data. In some cases, therefore, there was a need to rely on global sources of data. Details are provided in each of the supporting country reports (World Bank 2012a, 2012b, 2012c).

In addition, each country was divided into agricultural growing regions developed in collaboration with local experts. Areas within each region share similar characteristics in terms of terrain, climate, soil type, and water availability. As a result, baseline agricultural conditions, climate change impacts, and adaptive options are similar within each region, with some differences that are important for developing a specific adaptation plan.

Analytic Step 2: Develop Climate Projections

Climate change analyses require some forecasts of how temperature, precipitation, and other climate variables of interest might change over time. Because of the great uncertainty in climate forecasts, it is best in this type of study to attempt to characterize a range of alternatives as well as a "central case" forecast.

In this study the guiding principle used to select future climate scenarios was based on measures most likely to be relevant to negative or positive impacts of climate change on the agriculture sector. Because both temperature and precipitation affect agricultural productivity, scenarios were selected based on an index of soil moisture—the "climate moisture index" (CMI)—believed to be well correlated with potential agricultural production. The climate projections combine information from the baseline datasets with projections of changes in climate obtained from General Circulation Model (GCM) results prepared for the United Nations Intergovernmental Panel on Climate Change (IPCC) Fourth Assessment Report. (IPCC 2007).

As detailed in box 2.1, three climate scenarios were developed for each country, defined by the CMI, which measures the aridity of a region.[3] Using CMI values, the team selected for each country the driest, wettest, and a "medium" scenario from among 56 future climate change forecast scenarios developed by IPCC. Then both daily and monthly temperature and precipitation forecasts were generated to be used in the subsequent crop and water resources models.

Analytic Step 3: Conduct Impact Assessment

The goal of the impact assessment was to develop a rigorous quantitative assessment of the biophysical risks of climate change to agriculture if no adaptation were conducted. Subsequently the same model set was applied to estimate the marginal effect of individual adaptation measures on yields, which could then be valued and compared to the costs of those measures to assess the economics of alternative adaptation responses. As shown in figure 2.2, three general categories of biophysical models were used to develop the impact and adaptation assessments: crop models, a hydrological river runoff model, and a water balance model. The specific model choices within those categories were as follows:

- *Crop models.* Crop models analyze changes in crop yields and crop water and irrigation requirements. Different crop models were used in various combinations across the study countries (1) to assess which model could best provide a reasonable degree of confidence in the crop yield results; (2) to incorporate the effects of changes in temperature, precipitation, and irrigation water availability simultaneously; and (3) to be practically applied under multiple conditions to assess the marginal effect of individual adaptation measures needed to support B-C analyses. In prior work (Sutton, Srivastava, and Neumann 2013), the study team concluded that the Food and Agricultural Organization (FAO) AquaCrop model provided the best combination of high confidence in yield results, flexibility, and the ability to estimate marginal effects of adaptation measures; therefore the AquaCrop model was used here.

- *River runoff models.* These models are used to estimate the effects of climate change on the quantity of surface water available for irrigation and other uses.

Box 2.1 Developing a Range of Future Climate Change Scenarios

Analyzing climate change requires forecasting how temperature, precipitation, and other climate variables might change over time. The great uncertainty in these forecasts makes it necessary in a study like this to characterize a range of alternatives as well as a "central case" forecast. For temperature and precipitation projections, three climate scenarios were developed for the three countries: Low, Medium, and High Impact Scenarios.

Because both temperature and precipitation affect agricultural productivity, scenarios were selected based on a climate moisture index, or CMI. The CMI is based on the combined effect of temperature and precipitation, and as it is linked to soil moisture, so it is believed to be well correlated with potential agricultural production.

Each scenario in the study corresponds to a specific General Circulation Model (GCM) result combined with greenhouse gas (GHG) emissions scenarios. These SRES (*Special Report on Emissions Scenarios*) emissions scenarios were among those used by the Intergovernmental Panel on Climate Change (IPCC) in its fourth assessment of the science of climate change (IPCC 2000, 2007). The study relied on the three most commonly used GHG emissions scenarios: B1, A1b, and A2. As shown in table B2.1.1, a "wet" CMI scenario means that the location experienced the smallest impact (or change in) CMI—that is, the Low Impact Scenario. A dry scenario corresponds to high potential impact (High Impact Scenario). The Medium Impact Scenario reflects a central estimate of change in aridity. The specific global GCM selected for the medium scenario is closest in consistency with the model mean CMI from a total of 56 readily available GCM/SRES combinations.

The advantages of this approach are that it provides a representation of a full range of available scenarios for future climate change in a manageable way and that all climate scenarios are based on distinct GCM results. These results are themselves internally consistent in terms of the key GCM outputs the team used as inputs to the crop, livestock, and water resource impact modeling.

Table B2.1.1 Measurement Bases for Climate Impact Scenarios

Scenario	GCM model basis for the scenario	Relevant IPCC SRES scenario
Low Impact	National Center for Atmospheric Research, Parallel Climate Model (USA)	A2
High Impact	Goddard Institute for Space Studies, Model ER (USA)	A1B
Medium Impact	Center for Climate Modeling and Analysis, Coupled GCM 3.1 (Canada)	A1B

Note: SRES = Special Report on Emissions Scenarios (IPCC 2000).

Both temperature and precipitation changes affect river runoff volumes. The Climate and Runoff Hydrologic Model (CLIRUN) model was used to analyze changes in water runoff.

- *Water balance models.* These models combine information about the spatial layout of the water supply system with water demand and supply projections to assess whether certain uses might result in water shortages. Using the inputs from the river runoff model to characterize water supply, the crop modeling to characterize changes in irrigation water demand, and other analyses that project water demand from other users (such as hydropower and municipal water supply), this analysis used the water balance model primarily to identify potential shortages in water available for agriculture under climate change (Hughes, Chinowsky, and Strzępek 2010; Lehner et al. 2011; SEDAC 2011). The Water Evaluation and Planning System (WEAP) model was used in this analysis.

It is important to note that the analysis also included a critical "loop-back" from the results of the water balance modeling to the crop yield analysis, for any basin in which a water shortage for agricultural irrigation was noted (as illustrated in figure 2.2). The feedback loop was performed to estimate the yield of irrigated crops that might result if available water was insufficient for irrigation. The general increase in irrigation demands due to higher temperatures proved to be a very important part of the analysis.

The various modeling tools used in this analytic step are briefly described in box 2.2. If provided with less irrigation water than he or she demands, a farmer can either evenly distribute the remaining water over his cropland so that each crop receives less water (that is, deficit irrigation), or meet all the irrigation needs of a fraction of the crops, leaving the remaining fraction unwatered. The sensitivity of each crop planted to water shortages determines which approach will produce higher yields. For this important step in the analysis, information from FAO on the relationship between relative crop yield and relative water deficit—called the yield response factor (K_y)—was used to estimate the change in yield resulting from a reduction in water availability for each crop, relevant basin area, and climate scenario (FAO 1998).

Analytic Step 4: Evaluation of Adaptation Options

The adaptation options were evaluated primarily on the basis of five criteria: (1) net economic benefits (quantified, where possible, and otherwise based on expert assessment); (2) robustness to a range of potential climate scenarios; (3) potential to aid farmers with or without climate change, otherwise referred to as "win-win" potential; (4) favorable evaluation by stakeholders; and (5) potential for greenhouse gas (GHG) emission reductions. Because of data limitations, not all options are evaluated quantitatively. Methodologies for addressing each of the criteria are described as follows.

Criteria for Evaluting Adaptation Options

Criterion 1: Net Economic Benefits

Assessments of net economic benefits, conducted at the farm level on a per hectare basis, considered available estimates of the incremental cash costs for implementing the option, as well as the revenue implications of increasing crop yields. The net economic benefit model evaluates a subset of the adaptation options in

Box 2.2 Description of Modeling Tools for Impact Assessment

The three models used in this study are AquaCrop, Climate and Runoff Hydrologic Model (CLIRUN), and Water Evaluation and Planning System (WEAP). These models are in the public domain, have been applied world-wide frequently, and have a user-friendly interface. A brief description of each of these models follows.

AquaCrop

This model was developed and is maintained and supported by the Food and Agricultural Organization (FAO); it is the successor of the well-known CROPWAT package. The model is mainly parametric-oriented and therefore less data-demanding. It has the following strengths: (1) the simplicity to evaluate the impact of climate change and evaluation of adaptation strategies on crops and (2) the ability to evaluate the effects of water stress and estimate crop water demand. Figure B2.2.1 illustrates some of the main crop growth processes reflected in AquaCrop.

Figure B2.2.1 AquaCrop Model

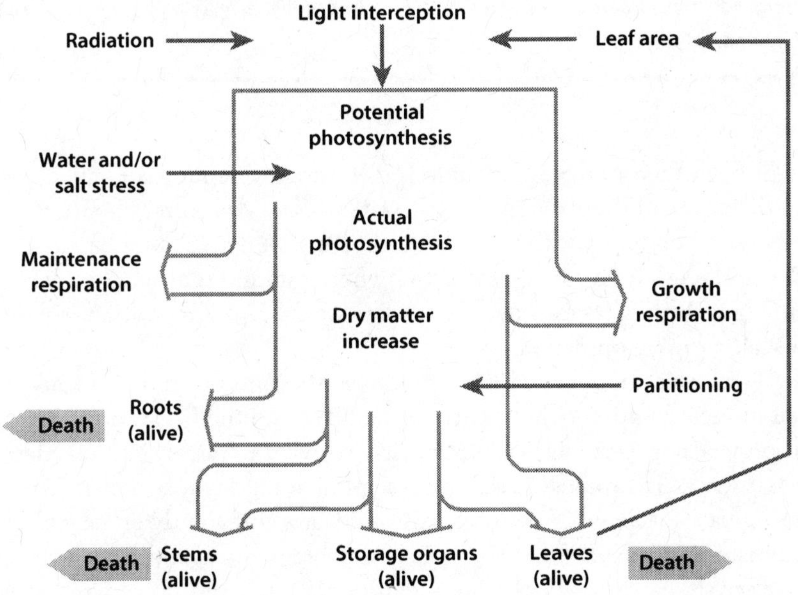

CLIRUN

The Climate and Runoff Hydrologic Model (CLIRUN) is widely used in climate change hydrologic assessments and can be parameterized using globally available data, but any local databases can also be used to enhance the data for modeling. It can run on a daily or monthly time step. CLIRUN can be used to estimate monthly runoff in a catchment. It models runoff as a lumped watershed with climate inputs and soil characteristics averaged over the watershed, simulating runoff at a gauged location at the mouth of the catchment. Soil water

box continues next page

Building Resilience to Climate Change in South Caucasus Agriculture
http://dx.doi.org/10.1596/978-1-4648-0214-0

Box 2.2 Description of Modeling Tools for Impact Assessment *(continued)*

is modeled as a two-layer system: a soil layer and groundwater layer. These two components correspond to a quick and a slow runoff response to effective precipitation. A suite of potential evapotranspiration (PET) models is also available for use in CLIRUN. Actual evapotranspiration is a function of potential and actual soil moisture states following the FAO method.

WEAP

WEAP—was developed by the Stockholm Environment Institute (SEI) and is maintained by the SEI U.S. Center. It is a software tool for integrated water resources planning that attempts to assist rather than substitute for the skilled planner. Although it is proprietary, SEI makes the model available for developing-country users. The software tool provides a comprehensive, flexible, user-friendly framework for planning and policy analysis. WEAP provides a mathematical representation of the river basin encompassing the configuration of the main rivers and their tributaries, the hydrology of the basin in space and time, and existing as well as potential major schemes and their various demands of water. The WEAP application used in the study models demands and storage in aggregate, providing a good base for future more-detailed modeling. For more information, see the *WEAP User Guide*, available at http://www.weap21 .org (Sieber and Purkey 2011).

---

terms of both their net present value (NPV—total discounted benefits less discounted costs) and their B-C ratio (B-C ratio—total discounted benefits divided by discounted costs) over the time period of the study. Ranking based solely on NPV would tend to favor projects with higher costs and returns, considering that the B-C ratio highlights the value of smaller scale adaptation options suitable for small-scale farming operations.

The economic model used here produces the optimal timing of adaptation project implementation by maximizing the NPV and the B-C ratio based on different project start years. This is particularly relevant to infrastructural adaptation options, such as irrigation systems and reservoir storage, whose high initial capital expenses may not be justified until crop yields are sufficiently enhanced. Finally, the model estimates NPV and B-C ratios for yield outputs under each dimension of the analysis, namely: (1) climate scenarios, (2) agricultural regions or (in the case of water supply options) river basins, (3) crops, (4) low and high agricultural commodity price forecasts, and (5) irrigated versus rainfed crops. Generating these metrics requires several key pieces of information, which include the following:

- *Crop yields* with and without the adaptation option in place, which are derived from the crop modeling. Changes in yields are modeled based on adaptations such as those that increase water availability, open irrigation in currently rainfed areas, optimize application of inputs, or result in more optimal use of crop varieties.[4]

Building Resilience to Climate Change in South Caucasus Agriculture
http://dx.doi.org/10.1596/978-1-4648-0214-0

- *Management multiplier* to convert from experimental to field yields: agronomic and crop modeling experts developed these estimates in consultation with local experts as part of their capacity-building work.
- *Crop prices* through 2050 were derived using national crop price data from FAO for current conditions and as a baseline to develop price projections under one scenario with constant prices and another based on the International Food Policy Research Institute (IFPRI) global price change forecast.
- *Exchange rates* between global and local crop prices were factored in.
- *Discount rate* to estimate the present value of future revenues and costs. The base case analyses employ a 5 percent discount rate consistent with recent World Bank economics of adaptation to climate change analyses (for example, World Bank 2013), but sensitivity tests using a 10 percent discount rate were also employed.
- *Capital and operations and maintenance (O&M) costs* of each adaptation input (for example, irrigation infrastructure). Local data were sought to characterize costs of adaptation options, and in some cases these data were provided. Overall, these can be difficult to obtain or generalize, and, as a result, in many cases estimates were derived from prior World Bank work or broader research.

The quantitative B-C analyses of adaptation options address in detail seven of the most important adaptation options as follows:

- Adding new irrigation capacity
- Rehabilitating existing irrigation infrastructure
- Improving water use efficiency in fields
- Adding new drainage capacity
- Rehabilitating existing drainage infrastructure
- Changing crop varieties
- Optimizing agronomic inputs (particularly fertilizer use)

Two of these options—improving water use efficiency and changing crop varieties—include costs for extension programs because extension must be enhanced to achieve the full benefits of the adaptation option. In addition, screening level analyses were conducted for four other options: expanding research and development, improving basin-level water use efficiency, adding new water storage capacity, and installing hail nets for selected crops.[5] These further analyses were more limited because of the lack of benefit information (requiring a "break-even" approach) or the inability to conduct the analysis at a crop-specific, model-farm level (for example, expanding research and development).

Criterion 2: Robustness to Different Future Climate Conditions

A key consideration in the quantitative analysis was assessing whether the option yields benefits across the range of possible future climate outcomes. These outcomes include quantitative and qualitative projections of net benefits of adaptation options across three climate change scenarios, two price scenarios, multiple crops, and four decades. All options were assessed relative to climate conditions in three alternative climate scenarios: Low, Medium, and High Impact. B-C ratios and NPV calculations were developed for each of the three scenarios, providing a means for assessing robustness to future climate conditions.[6]

Criterion 3: "Win-Win" Potential

The project team identified whether adaptation options would be beneficial, even in the absence of climate change. For options amenable to economic analysis, the team analyzed the net benefits of the adaptations relative to the current baseline; as a result, the benefits estimates implicitly incorporate both the climate adaptation and the non-climate-related benefits of adopting the measure. For other alternatives, the win-win potential was assessed based on expert judgment.

Criterion 4: Stakeholder Recommendations

Adaptation alternatives recommended by stakeholders during the stakeholder consultation workshops—at both the agricultural region and national levels— carried significant weight in the results. Stakeholders also provided information on impacts that they had already experienced and adaptation options that address those impacts. Adaptation options that addressed those impacts—even if those measures were not specifically mentioned in the stakeholder workshops— were also given a higher priority.

Criterion 5: Greenhouse Gas Mitigation Potential

Once an initial set of options was identified as high priority, the team then also analyzed the GHG mitigation potential of adaptation options. For this study, adaptation effectiveness for agriculture was the highest priority criterion, with GHG mitigation potential identified as an ancillary benefit once the option was established as cost-effective, highly desired by stakeholders, or possessing "win-win" potential.

Limitations and Key Challenges

While the approaches developed and applied in this assessment need to be as robust and accurate as possible, they must also reflect local data availability and must avoid unnecessary complexity to achieve the goals of in-country capacity-building and stakeholder involvement. The framework was designed to be suitable for a wide range of crops (for example, maize, wheat, tomato, wine grapes, apple, alfalfa, and cotton) selected for focus in the early stages of each country analysis. The resulting methodology is suitable to simulate and evaluate a range of adaptation options for various climate change scenarios, cropping systems, and agricultural water regimes.

A study with so broad a scope necessarily has significant limitations. For example, assumptions must be made about many important aspects of agricultural and livestock production in each country, the limits of simulation modeling techniques for forecasting crop yields and water resources must be considered, and time and resource constraints must be factored in. The overall methodology was designed to yield results sufficiently precise to ensure that the adaptation measures will yield benefits in excess of costs and are robust to future climate change. Some of the options will require additional, more detailed examination and analysis to ensure that specific adaptation measures are implemented in a manner that maximizes their value to agriculture in each country.

Nevertheless, while more detailed modeling could yield more precise impact and B-C results, pursuing a more detailed approach would not necessarily alter the ranking of options or suggest that options evaluated to be highly cost-effective might instead be poor investments. In order to look broadly across many crops, areas, and adaptation options, however—particularly for adaptation options that may be relatively new to each of the countries supported in the study—it was necessary to develop general data and characterizations of these options. While the study team took great care to use the best available data and applied state-of-the-art modeling and analytic tools, they recognized that analysis of outcomes 40 years into the future, across a broad and varied landscape of complex agricultural and water resources systems, involves uncertainty. As a result, the team attempted to evaluate the sensitivity of results to one of the most important sources of uncertainty—how future climate change will unfold—through the use of the multiple climate scenarios.[7]

Other costs and benefits that do not affect farm expenditures or revenues are excluded from the quantitative analysis, mainly owing to the lack of available data. For example, while increasing fertilizer use may lead to social costs in terms of negative effects on nearby water quality, it is very difficult to quantify those effects without consideration of the site-specific characteristics that may be unique to individual farms.

A potentially larger question, more difficult to address, involves projecting the evolution and development of agricultural systems over the next 40 years, with or without climate change. The future context in which adaptation will be adopted is clearly important but very difficult to forecast. Other important limitations involve the necessity of examining the efficacy of adaptation options for a "representative farm." The result is an important initial step in the process of evaluating and implementing climate adaptation options for the agriculture sector using the current best available methods.

The researchers hope, however, that the awareness of climate risks and the analytic capacities built through the course of this study provide not only a greater understanding among agricultural institutions of the basis of the results, but also an enhanced capability to conduct the more detailed assessment that will be needed to further pursue the most promising adaptation measures.

Notes

1. In this analysis, the economic and physical baseline is current yields, which represents a simplification of the expectation for these countries but is a reasonable expectation for agricultural productivity without planned adaptation interventions. Because the

purpose of this study is to evaluate measures that might enhance resilience to both current and future climate, it is not clear whether modeling of an alternative baseline that includes agricultural innovation (and adoption) is appropriate or important. Using a baseline of increasing yield, for example, implies that some adaptive actions (such as new varieties) would be adopted as "autonomous" adaptations, at some cost to either the country or the farmers. This study examines marginal gains in crop yields and farm-level revenue from this baseline for individual measures that the team believes are unlikely to be adopted without additional adaptation plans and investments. It is certain that projecting a baseline of future crop yields that differs from the constant yield assumption used here adds significant complexity and uncertainty to the results.

2. Some might argue that simulation modeling is so demanding of inputs that it yields less precise or even inaccurate estimates. The difficulties of simulation modeling make calibration of the models to current conditions, wherever possible, most important. The Ricardian approach, an econometric evaluation of historical agricultural sector performance, is sometimes put forward as an alternative method for estimating the impacts of climate change on yields and revenue in response to climate change. However, the Ricardian approach, which relies on an econometric estimation of a climate response function based on current data, implicitly reflects adaptive responses in the current system and therefore lacks the ability to estimate the incremental benefits of specific adaptation options. Furthermore, only currently practiced adaptive measures are reflected in the estimation—whereas in many cases in developing countries agricultural systems are poorly adapted to current climate, reflecting an "adaptation deficit"—and new measures should be introduced.

3. The CMI depends on average annual precipitation and average annual potential evapotranspiration (PET). If PET is greater than precipitation, the climate is considered to be dry, whereas if precipitation is greater than PET, the climate is moist. Calculated as CMI = (P/PET) – 1 {when PET > P}; and CMI = 1 – (PET/P) {when P > PET}; a CMI of –1 is very arid and a CMI of +1 is very humid. As a ratio of two depth measurements, CMI is dimensionless.

4. For changes in varieties, the team looked not at the yield benefits of newly developed seed varieties, but rather at adopting currently available varieties that are either not used at present or that would optimize yields for future conditions. A separate analysis reviews possible returns from investment in research to develop new varieties and technologies.

5. Note that some analysts have suggested that improving water use efficiencies, such as lining irrigation channels, may have little value if both surface water and groundwater are used for irrigation, because losses from the channels would be gains to the groundwater aquifers. However, the cost of collecting and delivering the water to the fields must be taken into account, so while the water may not be lost to the hydrologic system, additional pumping costs would be incurred to recover water lost from irrigation channels.

6. An interesting finding is that in most cases quantitative results for adaptation options were less sensitive to uncertainties in climate forecasts than to uncertainties in future prices. This was also true for CO_2 fertilization effects on yield.

7. The following chapters, which show the climate projections for each country, demonstrate that using multiple climate scenarios is a critical step and that use of only one scenario would suggest more certainty in climate forecasts than is warranted, particularly for precipitation projections that are critical for agriculture.

References

FAO (Food and Agriculture Organization of the United Nations). 1998. *Crop Evapotranspiration: Guidelines for Computing Crop Water Requirements*. FAO Irrigation and Drainage Paper 56, Rome: FAO (accessed December 9, 2013), http://www.fao.org/docrep/x0490e/x0490e00.htm#Contents.

Hughes, G., P. Chinowsky, and K. Strzępek. 2010. "The Costs of Adaptation to Climate Change for Water Infrastructure in OECD countries." *Utilities Policy* 18 (3): 142–53.

IPCC (Intergovernmental Panel on Climate Change). 2000. *Special Report on Emissions Scenarios: A Special Report of Working Group III of the IPCC*, edited by N. Nakicenovic and R. Swart. Cambridge, U.K.: Cambridge University Press.

———. 2007. *Contribution of Working Group I to the Fourth Assessment Report of the Intergovernmental Panel on Climate Change*, edited by S. Solomon, D. Qin, M. Manning, Z. Chen, M. Marquis, K. B. Averyt, M. Tignor, and H. L. Miller. New York: Cambridge University Press.

Lehner, B., C. Reidy Liermann, C. Revenga, C. Vorosmarty, B. Fekete, P. Crouzet, P. Doll, M. Endejan, K. Frenken, J. Magome, C. Nilsson, J.C. Robertson, R. Rodel, N. Sindorf, and D. Wisser. 2011. Global Reservoir and Dam Database, Version 1 (GRanDv1): Dams, Revision 01. Palisades, NASA Socioeconomic Data and Applications Center (SEDAC), New York (accessed October 10, 2013), http://sedac.ciesin.columbia.edu/data/set/grand-v1-dams-rev01.

SEDAC (Socioeconomic Data and Applications Center). 2011. "Gridded Population of the World." Columbia University, New York (accessed January 15, 2011), http://sedac.ciesin.columbia.edu/gpw/.

Sieber, J., and D. Purkey. 2011. *Water Evaluation and Planning System User Guide*. Somerville, MA: Stockholm Environment Institute.

Sutton, W., J. Srivastava, and J. Neumann. 2013. *Looking Beyond the Horizon: How Climate Change Impacts and Adaptation Responses Will Reshape Agriculture in Eastern Europe and Central Asia*. Washington, DC: World Bank.

World Bank. 2012a. "The Republic of Armenia: Climate Change and Agriculture Country Note—June 2012." World Bank, Washington, DC (accessed January 17, 2014), http://siteresources.worldbank.org/ARMENIAEXTN/Resources/CN_Armenia_FINAL.pdf.

———. 2012b. "The Republic of Azerbaijan: Climate Change and Agriculture Country Note—June 2012." World Bank, Washington, DC (accessed January 17, 2014), http://siteresources.worldbank.org/AZERBAIJANEXTN/Resources/CN_Azerbaijan_FINAL.pdf.

———. 2012c. "Georgia: Climate Change and Agriculture Country Note—June 2012." World Bank, Washington, DC (accessed January 17, 2014), http://siteresources.worldbank.org/GEORGIAEXTN/Resources/CN_Georgia_FINAL.pdf.

———. 2013. "Armenia, Azerbaijan and Georgia: Where Climate-Resilient Agriculture Means Less Poverty." World Bank, Washington, DC (accessed December 9, 2013), http://go.worldbank.org/1A06SPQM50.

CHAPTER 3

Armenia: Risks, Impacts, and Adaptation Menu

This chapter summarizes the results of efforts to develop a menu of adaptation options for the agricultural sector in Armenia. It is organized into four sections: (1) climate risk, (2) climate impacts, (3) adaptation assessment, and (4) evaluation and prioritization of adaptation options.

Climate Risk

Historical Climate Trends

The South Caucasus region has seen a variety of changes in climate, including increasing temperatures, shrinking glaciers, sea level rise, reduction and redistribution of river flows, decreasing snowfall, and an upward shift of the snowline. In the past 10 years, the region has also experienced more extreme weather events—flooding, landslides, forest fires, and coastal erosion—resulting in economic losses and human casualties (WWF 2009).

Figures 3.1 and 3.2 present historical temperature and precipitation data for Armenia. Figure 3.1 shows annual temperatures and growing season temperatures, 1900–2012. During 1980–2012, average annual temperature and average growing season temperature both increased by approximately 1°C.

Figure 3.2 presents average monthly precipitation over the year and average growing season precipitation, 1900–2012. During 1980–2012, the average monthly precipitation increased approximately 11.5 mm, while average growing season precipitation increased approximately 19.3 mm.

In addition to the temperature and precipitation changes, the glaciers are melting rapidly in the region, as they are globally. The volume of glaciers in the South Caucasus has been reduced by 50 percent over the last century, and 94 percent of the glaciers have retreated 38 meters per year (Stokes et al. 2006). Changes in glacier composition can potentially reduce long-term river flow in Armenia.

Figure 3.1 Average Annual and Growing Season Temperatures in Armenia, 1900–2012

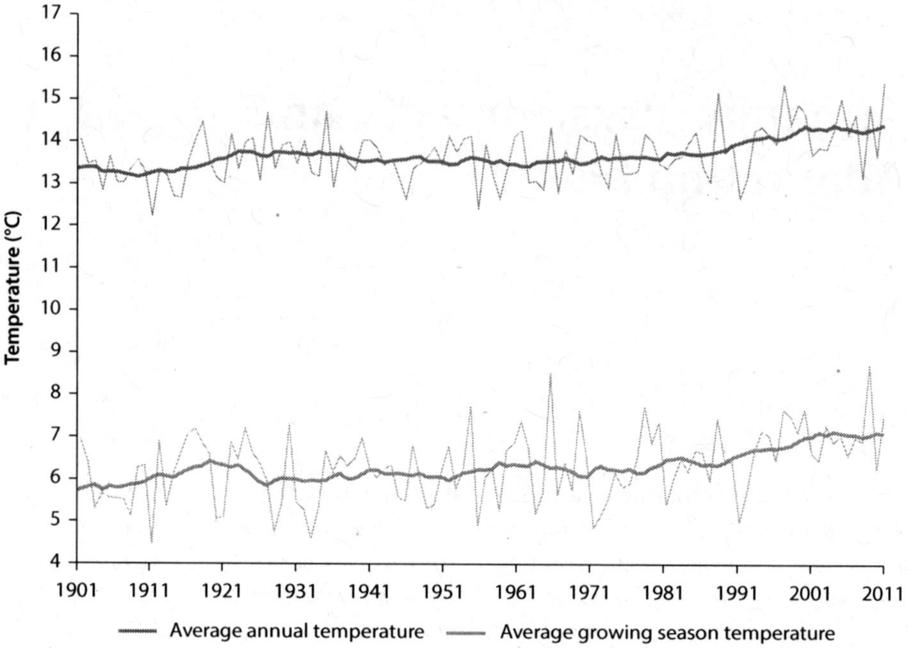

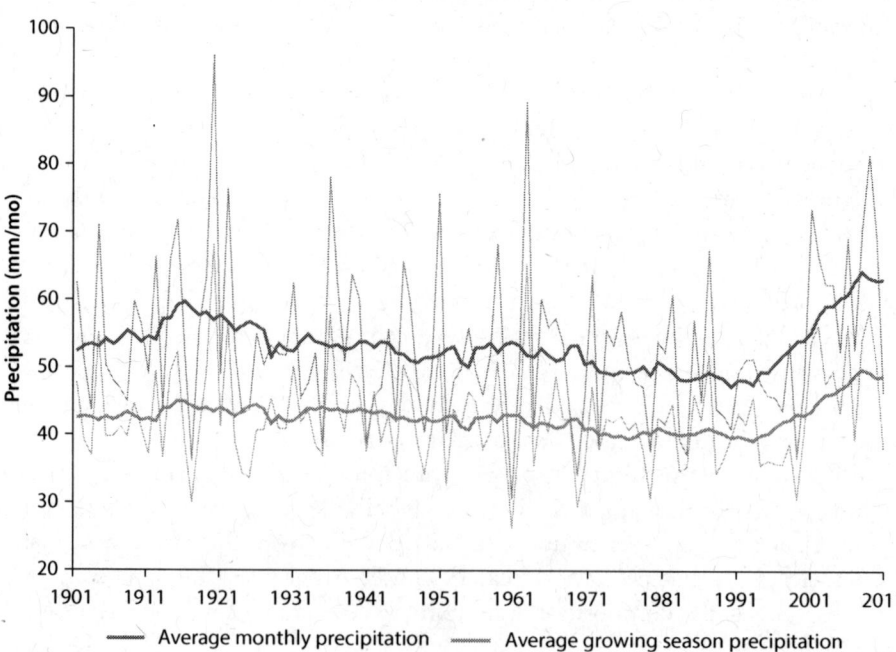

Source: University of East Anglia Climatic Research Unit, Norwich, UK.

Figure 3.2 Average Monthly and Growing Season Precipitation in Armenia, 1900–2012

Source: University of East Anglia Climatic Research Unit, Norwich, UK.
Note: mm/mo = millimeters per month.

Forecasted Changes in Temperature and Precipitation

Analyses of recent climate data and information gathered from the study's farmer workshops support the study finding of a trend of increasing temperature in Armenia, and also reveal that the frequency of extreme temperature events is also increasing in the country. The results of this study indicate that this warming trend will accelerate in Armenia in coming decades, as shown on map 3.1. Although the degree of warming that will occur in Armenia remains uncertain, the overall warming trend is clear and evident in all three of Armenia's agricultural regions—mountainous, intermediate, and lowlands—with average warming over the next 50 years for the Medium Impact Scenario estimated at about 2.6°C, much greater than the increase of less than 0.85°C observed over the last 80 years (UNFCCC 2010). Warming could be more modest, but average temperature changes for the Low Impact Scenario nonetheless represent an increase of about 1.2°C, compared to current conditions.

Changes in precipitation are harder to predict, and estimates of how precipitation will change in Armenia are uncertain, as shown on map 3.2. Under the Medium Impact Scenario, nationwide precipitation decreases approximately 52 millimeters (mm) per year on average by the 2040s. However, the range of precipitation outcomes across the Low and High Impact alternatives is large, ranging from a modest increase under the Low Impact Scenario to a 19–28 percent decline under the High Impact Scenario. Uncertainty at the regional level is even higher, and annual precipitation declines in the highest elevation agricultural region could be as large as 144 mm per year.

Climate change may also increase the frequency and magnitude of droughts, frosts, and floods in Armenia. While precipitation is expected to increase only under the Low Impact Scenario by the 2040s (map 3.2), rainfall events are predicted to become more variable, with a high probability of daily to multi-day events becoming larger and less frequent. Such flood events pose a particular threat to the agriculture sector in Armenia in the spring, when flooding can delay or prevent planting of summer crops, and during late summer, when flooding can destroy an entire year's growth and prevent timely harvesting. Even small flood events can reduce productivity, since prolonged water-logging is detrimental to many crops.

Finally, the yearly averages of temperature and precipitation are less important for agricultural production than are the seasonal distribution of temperature and precipitation. Under climate change, temperature increases are predicted to be highest in the period July–October relative to current conditions. This summer temperature increase can be as much as 5°C in the intermediate agricultural region of Armenia. In addition, forecasted precipitation declines are greatest in the key July–August period, when precipitation is already near its lowest. Figure 3.3 presents the monthly baseline and forecasted temperatures and precipitation changes for the intermediate agricultural region.

Building Resilience to Climate Change in South Caucasus Agriculture
http://dx.doi.org/10.1596/978-1-4648-0214-0

Map 3.1   Armenia: Predicted Effect of Climate Change on Average Annual Temperature in the 2040s

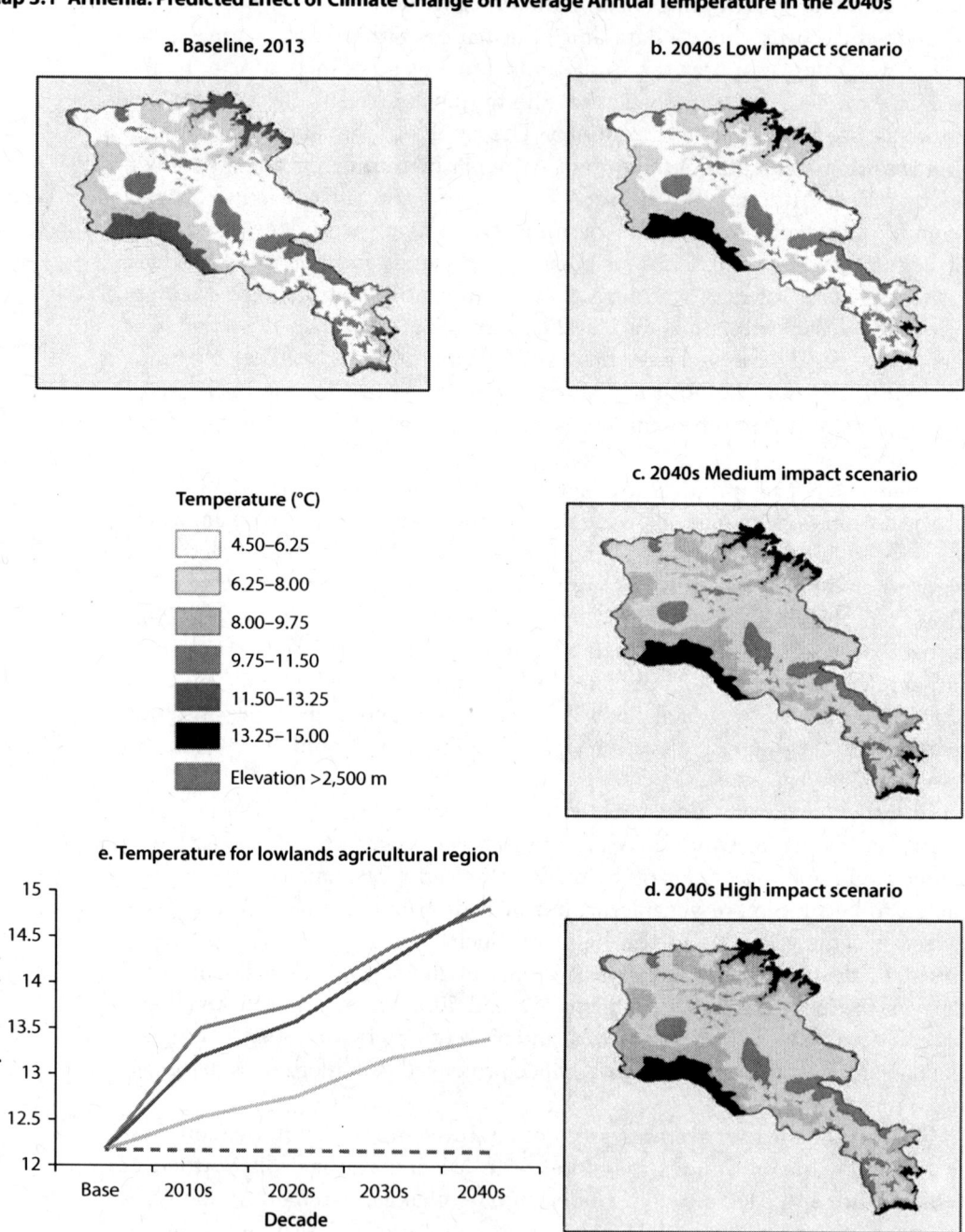

a. Baseline, 2013

b. 2040s Low impact scenario

c. 2040s Medium impact scenario

Temperature (°C)

4.50–6.25
6.25–8.00
8.00–9.75
9.75–11.50
11.50–13.25
13.25–15.00
Elevation >2,500 m

e. Temperature for lowlands agricultural region

d. 2040s High impact scenario

High   Medium   Low   Base

Sources: ©Industrial Economics. Used with permission; reuse allowed via Creative Commons Attribution 3.0 Unported license (CC BY 3.0). Country boundaries are from ESRI and used via CC BY 3.0.

Map 3.2 Armenia: Predicted Effect of Climate Change on Average Annual Precipitation in the 2040s

a. Baseline, 2013

b. 2040s Low impact scenario

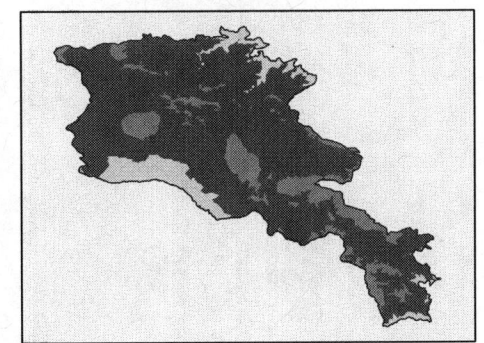

Precipitation (mm/yr)

- 275–325
- 325–375
- 375–425
- 425–475
- 475–525
- 525–575
- Elevation >2,500 m

c. 2040s Medium impact scenario

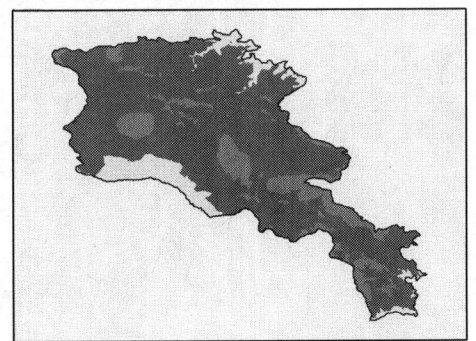

e. Precipitation for lowlands agricultural region

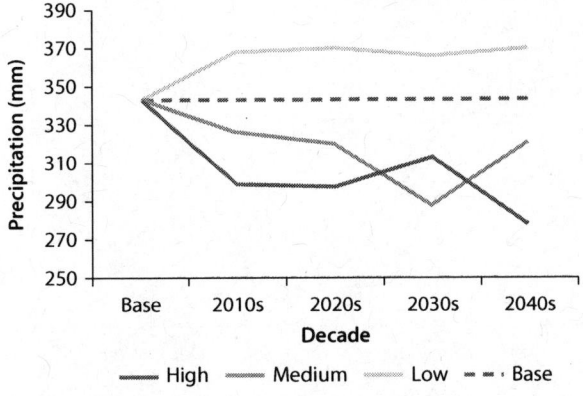

d. 2040s High impact scenario

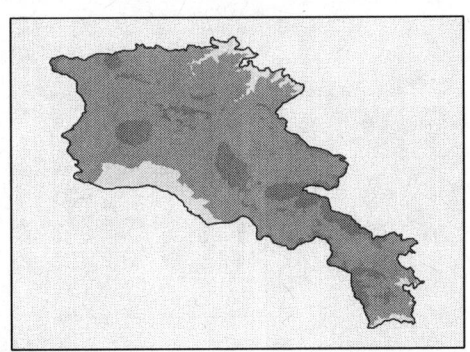

Sources: ©Industrial Economics. Used with permission; reuse allowed via Creative Commons Attribution 3.0 Unported license (CC BY 3.0). Country boundaries are from ESRI and used via CC BY 3.0.
Note: mm = millimeters.

Figure 3.3 Armenia: Effect of Climate Change on Monthly Temperature and Precipitation Patterns for the Intermediate Agricultural Region in the 2040s

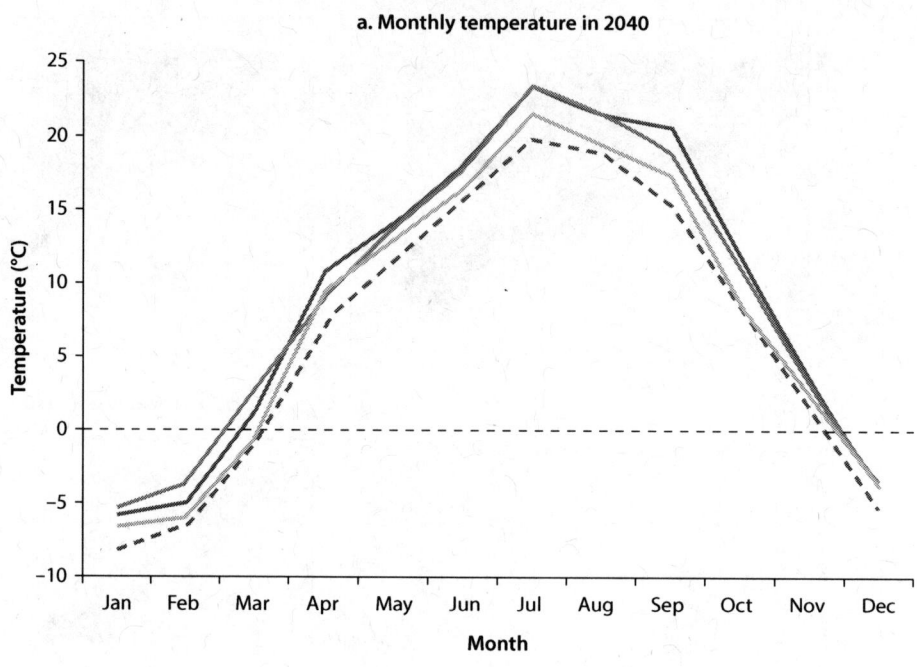

a. Monthly temperature in 2040

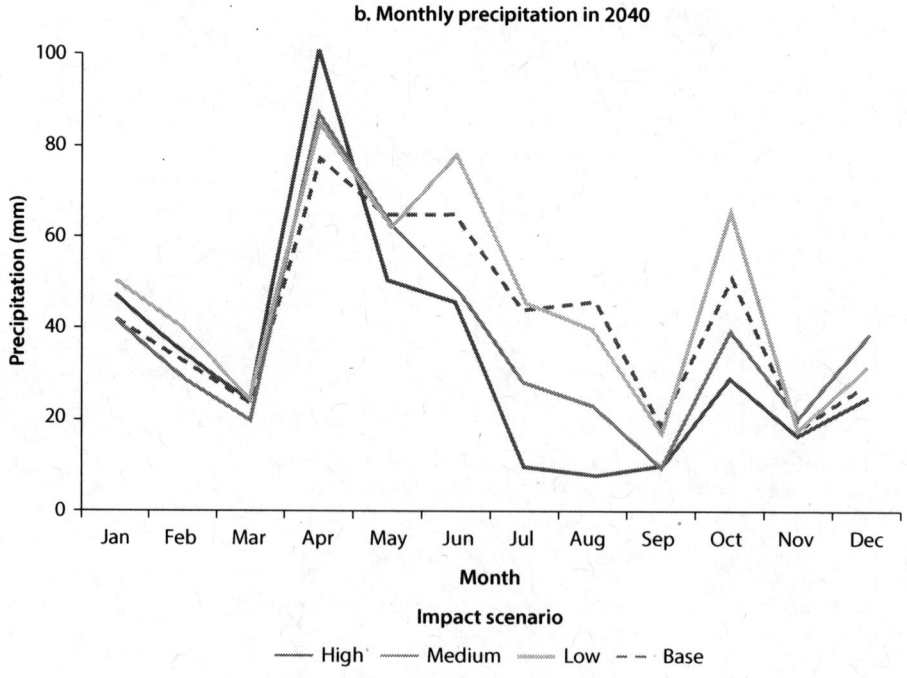

b. Monthly precipitation in 2040

Impact scenario

——— High ———— Medium ———— Low – – Base

Source: World Bank data.
Note: mm = millimeters.

Climate Impacts

In order to assess the impact of climate change on the agricultural sector in Armenia, the monthly projections of temperature and precipitation were translated to daily projections for use in crop models, as described in chapter 2, box 2.2. The crop models examined the potential effect of climate change on crop yields in Armenia under the "no adaptation" scenario (that is, if no adaptation measures are taken). The crop yield impacts presented in table 3.1 represent the potential outcome under the Medium Impact Scenario and do not take into account irrigation water constraints.

Decline in Crop Yields

As shown in table 3.1, yields of alfalfa, apricot, grape, and potato are expected to decline across all agricultural regions in the 2040s under the Medium Impact Scenario. Yields of wheat, Armenia's key cereal crop, are expected to increase in the mountainous and intermediate regions, but decrease in the lowlands region due to rising temperatures and water stress. Tomato yields are also expected to increase in the mountainous and intermediate agricultural regions, and irrigated watermelon yields are expected to increase in the intermediate region.

Although table 3.1 reflects the assumption that irrigation water will not be constrained, changes in temperature and precipitation resulting from climate change are expected to impact water resources in Armenia. As a result, a more detailed water resource analysis is also needed to determine the extent of

Table 3.1 Effect of Climate Change on Crop Yields in the 2040s under the Medium Impact Scenario, No Adaptation and No Irrigation Water Constraints

Irrigated/rainfed	Crop	Change in yield (%)		
		Lowlands	Intermediate	Mountainous
Irrigated	Alfalfa	−5	−7	−2
	Apricot	−5	−5	−5
	Grape	−7	−5	−5
	Potato	−12	−9	−5
	Tomato	−16	6	50
	Watermelon	−12	10	n.a.
	Wheat	−6	1	38
Rainfed	Alfalfa	−3	−8	−1
	Apricot	−28	−7	−5
	Grape	−24	−12	−1
	Potato	−14	−14	−8
	Tomato	−19	−8	34
	Watermelon	−18	0	n.a.
	Wheat	−8	1	38

Source: World Bank data.
Note: Results are average changes in crop yield, assuming no effect of carbon dioxide fertilization, under Medium Impact Scenario (no adaptation and no irrigation water constraints). Declines in yield are shown in shades of orange, with darkest representing biggest declines; increases are shaded green, with darkest representing the biggest increases.
n.a. = not applicable (indicates that the crop was not analyzed in that country).

climate change impacts. The study team conducted a water availability analysis for Armenia using the Water Evaluation and Planning System (WEAP) and the Climate and Runoff Hydrologic Model (CLIRUN) model. Next, water supply estimates were matched with forecasts of water demand for all sectors, including agriculture, to determine water availability. Agricultural water demand was estimated using the AquaCrop model (see chapter 2, box 2.2 for more information).

Water Supply Declines, Demand Increases

Figure 3.4 presents the estimated effect of climate change on mean monthly runoff in Armenia in the 2040s. The runoff indicator is directly relevant to agriculture systems and provides insight into the risk of climate change for agricultural water availability, as well as the implications of climate change for water resource management. As shown in figure 3.4, under the High Impact Scenario, overall water supply is expected to decline by an average of 30–40 percent by the 2040s. At the same time, irrigation water demand during the summer months is expected to increase by up to 20 percent relative to historic demands. The net effect of the predicted rising demands and falling supply is a significant reduction in water available for irrigation. Irrigation water shortages by the 2040s are predicted to occur in the Upper Araks basin, while no shortage of irrigation water is forecasted for the other Armenian basins.

Figure 3.4 Estimated Effect of Climate Change on Mean Monthly Runoff in the 2040s for All Armenian Basins

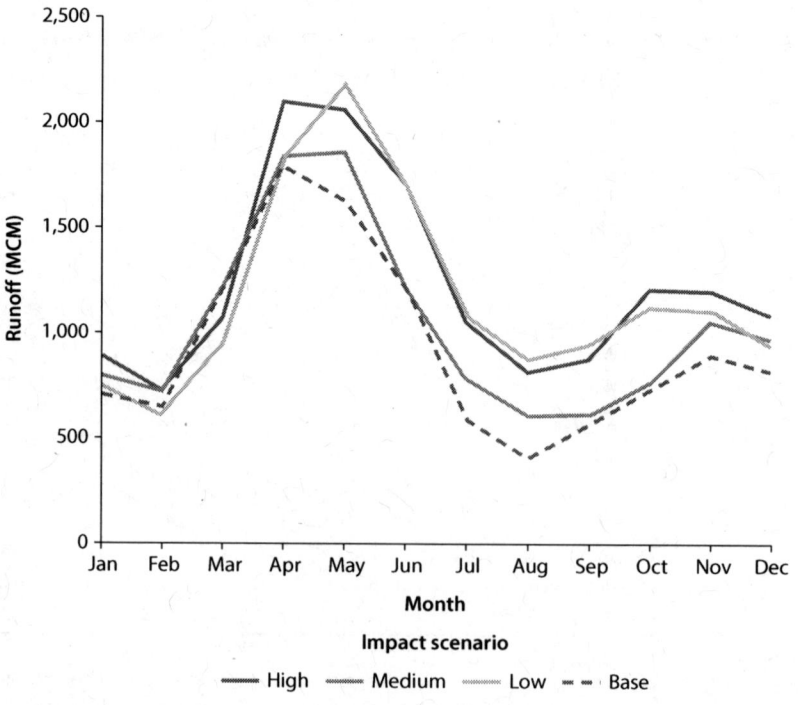

Source: World Bank data.
Note: MCM = million cubic meters.

Negative Net Climate Effects

Therefore three climate change stressors combine to yield an overall negative impact on crop yields in Armenia:

- The direct effect of temperature and precipitation changes on crops
- Increased irrigation demand required to maintain yields
- Decline in water supply associated with higher evaporation and lower rainfall

All of these effects have a more pronounced impact during the summer growing season. For example, even though annual runoff is forecasted to increase under the Low Impact Scenario, it is expected to decline during the late spring and late summer months under all three scenarios relative to baseline conditions, which is exactly when irrigation water demand is highest. The net effect of these three factors on irrigated agriculture is illustrated in table 3.2.

The study analysis reveals that in Armenia the main effect of climate change on availability of agricultural water (which results from the combined effect of temperature and precipitation changes and decline in water supply) will be on

Table 3.2 Effect of Climate Change on Irrigated Crop Yields Adjusted for Estimated Irrigation Water Deficits in the Upper Araks Basin in the 2040s

a. Crop yield impacts due to temperature and precipitation changes without considering irrigation water constraints

Crop	Change in yield (%)		
	Lowlands	Intermediate	Mountainous
Alfalfa	−5	−7	−2
Apricot	−5	−5	−5
Grape	−7	−5	−5
Potato	−12	−9	−5
Tomato	−16	6	50
Watermelon	−12	10	50
Wheat	−6	1	38

b. Crop yield impacts due to temperature and precipitation changes as well as forecasted irrigation water constraints

Crop	Change in yield (%)		
	Lowlands	Intermediate	Mountainous
Alfalfa	−48	−49	−46
Apricot	−48	−47	−47
Grape	−42	−41	−41
Potato	−51	−49	−47
Tomato	−53	−41	−17
Watermelon	−51	−39	−17
Wheat	−48	−44	−24

Source: World Bank data.
Note: Results are average changes in crop yield, assuming no effect of carbon dioxide fertilization. Declines in yield are shown in shades of orange, with darkest representing biggest declines; increases are shaded green, with darkest representing the biggest increases.

the Upper Araks basin, which feeds the Ararat Valley. The net effect of the three factors on irrigated agriculture in the Upper Araks basin is illustrated in table 3.2. Table 3.2a shows the effect of temperature and precipitation changes alone on irrigated agriculture if there are no irrigation water constraints. Table 3.2b shows the combined effect of all three factors mentioned above, including the forecasted irrigation water shortages for the Upper Araks basin. The net effect of these factors on crop yields is dramatic, and provides an important focus for adaptation efforts to mitigate potential losses. While the water resources modeling does not indicate water shortages for the Lower Araks basin, changes in transboundary water withdrawal rates could alter that finding and lead to shortages in that part of the Araks basin as well.

The direct effects of climate change on livestock production in Armenia could also be severe, but the methods available for quantitatively assessing effects on livestock are relatively untested. There is a robust literature establishing that increases in temperature decrease livestock productivity (Thornton et al. 2009), but suitable modeling tools for quantifying the effect in the Armenian context are not available. According to the analysis in this study, the indirect effect of climate change on livestock feedstocks including pasture would be positive, thus providing a counterbalance to the negative direct heat stress effects cited in the literature.

Adaptation Assessment

After examining the local climate risk and likely impacts of climate change on Armenia's agricultural sector, the study team conducted an adaptation assessment of the sector, both at the national and regional levels. This involved stakeholder outreach to elicit information about current farming practices, observed impacts of climate change thus far, and how farmers are currently adapting to these impacts. In addition, the stakeholder outreach sessions allowed the study team to compile an initial list of priority adaptation options based on input from farmers as well as government officials and other local experts. This section describes the findings of the adaptation assessment and the recommended adaptation options from the stakeholder consultations.

Current Regional Adaptive Capacity

To assess Armenia's current regional adaptive capacity, it was essential that the study team inform and consult with a variety of local stakeholders—farmers and farmers' associations, local government officials, students studying agriculture, and other local experts—on the predicted impacts of climate change on agriculture and water resources. The team first met with farmers for a one-day stakeholder workshop in Yeghegnadzor in April 2012. A second set of farmer consultations was conducted in October 2012 at three locations (Martuni, Artashat, and Yeghegnadzor), representing the different agricultural regions of Armenia (map 3.3).

At the initial workshop, participants were given an overview of the study and the potential impacts of climate change on crop yields and water

Map 3.3 Locations of Stakeholder Consultations in Armenia

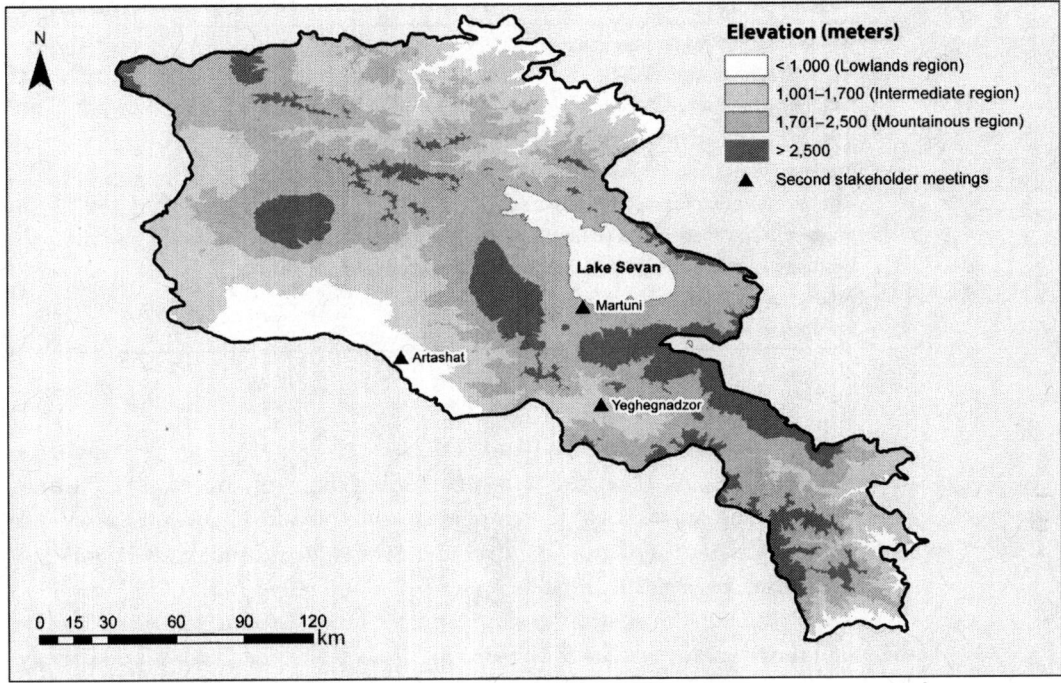

Sources: ©Industrial Economics. Used with permission; reuse allowed via Creative Commons Attribution 3.0 Unported license (CC BY 3.0). Country boundaries are from ESRI and used via CC BY 3.0.
Note: km = kilometers.

availability in Armenia. They were then asked if they had witnessed climate change impacts and what they have done, or would do, to mitigate their effects. All confirmed that several of the impacts have been felt on local farms. The stakeholders at the workshop made it clear that, although farmers are becoming more flexible in their response to climate events, their adaptive capacity is still quite limited due to poorly maintained irrigation and drainage systems, limited financial resources, and inadequate support from and access to extension services.

At the subsequent farmer consultations, participants were provided with a list of potential climate adaptations. They were asked to remove any irrelevant adaptations and add any additional adaptations that they believed would be effective. Participants then provided rankings for both national-level and regional-level adaptation options. Rankings of regional-level options varied among the regions, reflecting differences in current climates, topography, and other location-specific factors. The ranked recommendations of adaptation options for each of Armenia's three agricultural regions are as follows.

Lowlands Agricultural Region: Artashat

The agricultural sector in this region produces a variety of crops, including wheat, vegetables, watermelons, grapes, and orchard fruits, as well as livestock.

Building Resilience to Climate Change in South Caucasus Agriculture
http://dx.doi.org/10.1596/978-1-4648-0214-0

Table 3.3 Ranked Recommendations from the Artashat Consultation

Adaptation option	Points
Rehabilitation of water reservoirs	26
Rehabilitation of irrigation	25
Optimize application of water	20
Reduce erosion and soil conservation	15
Improve livestock nutrition and shelter	9
Optimize agronomic practices (fertilizer)	9
Improve crop varieties, particularly those tolerant to droughts	9
Restoration of pastures by improved agronomic practices	7
Adjust type of crops based on elevation	6
Hail rockets	4

The region's climate is sufficiently mild that two crops a year can be grown, but farmers find it necessary to use irrigation rather than rely on rainfall. Farmers reported noticing an increase in temperature in this already warm climate, in addition to a greater frequency of extreme events such as drought, hail, and heat waves, resulting in negative impacts on crops.

The importance of irrigation to support agricultural production is apparent in the adaptation rankings (table 3.3). Farmers stressed the need for adequate irrigation water to ensure both quantity and quality of orchard and vineyard production. In addition, the rankings support the fact that livestock are an important part of the agricultural economy as they can beneficially use field crop aftermath (second growth) as well as rainfed rangeland. Improved livestock husbandry and health and optimizing the production and storage of livestock forage were aggregated as a single measure and ranked fifth, tied with improved crop production practices and improved crop/livestock genetics.

Local orchardists reported some innovative attempts to reduce climatic risk by interplanting crops with different climate sensitivities. Examples are an apricot orchard with a peach tree planted as every other tree to hedge against early spring frosts that might damage apricots but not the later-flowering peaches, and a vineyard with tomato planted in between the rows of vines for the same reason.

Intermediate Agricultural Region: Yeghegnadzor

Farmers in this region reported that the climate was becoming warmer and that extreme weather events were more frequent. They noted that the most important weather-related impact is drought, which is especially burdensome due to its variability and extremes. Changes in the cropping season, hail, winter frost, warming, and increasing water demand also negatively affect crop production in this region of Armenia. With the crop seasons shifting, farmers plant earlier, but spring freezing can harm crops. Hail has also worsened recently, especially in the spring when it hits early vegetation. Winter frost was noted, especially during the winter of 2002 when trees were completely frozen. Increasing temperatures have resulted in increased incidences of diseases, pests, and weeds, as well as

Table 3.4 Ranked Recommendations from the Yeghegnadzor Consultation

Adaptation option	Points
Rehabilitation of irrigation	26
Adjust type of crops based on elevation	23
Optimize agronomic practices (fertilizer)	15
Improve crop varieties, particularly those tolerant to droughts	13
Reduce erosion and soil conservation	12
Improve livestock nutrition and shelter	11
Hail rockets	8
Optimize application of water	8
Restoration of pastures by improved agronomic practices	6
Rehabilitation of water reservoirs	3

the emergence of new types of pests and diseases. Finally, crop water demand continues to increase, which can become problematic socially as people have to pay more for water.

Generally, farmers have observed the changing climate and have already begun responding. Many are planting crops earlier to respond to higher temperatures earlier in the season, moving their crops to higher elevation areas, changing crop rotations, and changing the timing of irrigation. Highly ranked adaptation options (table 3.4) include rehabilitation of aging irrigation systems and relocating orchards to less frost-prone sites, as well as application of a variety of other basic improved practices dealing with crop and livestock production.

Mountainous Agricultural Region: Martuni

Farmers in this region rely on irrigation for crop production, with non-irrigated land often used as unimproved pasture. Major crops include wheat, potato, and cabbage. The major climatic changes noted were increased temperatures, more frequent heat waves, and droughts. Farmers reported that disease and pest problems were also increasing, perhaps as a byproduct of climate change, and that these have resulted in crop damage. The high rankings given to irrigation-related adaptations (table 3.5) clearly reflect the importance of irrigation to crop and fruit production in this region. Farmers raise livestock but have limited pasture to support them and are aware of the need to improve basic animal husbandry practices.

Current National-Level Adaptive Capacity and Responses

Participants in all three regions generally agreed about the need for low-interest, long-term loans for farmers to help them implement adaptation measures. This recommendation, along with crop insurance, was by far the highest-ranked item of the adaptation options (table 3.6). Currently it is difficult for farmers to obtain loans, and those available are most often short-term and high-interest. Farmers reported that although crop insurance was sometimes available from the private market, it is often too expensive. They were very interested in securing insurance against losses such as hail and frost. The second tier of adaptation options reflects

Table 3.5 Ranked Recommendations from the Martuni Consultation

Adaptation option	Points
Rehabilitate irrigation systems	24
Construct small volume reservoirs	19
Provision of agricultural equipment	19
Improve crop varieties	9
Improve livestock nutrition and shelter	7
Optimize application of irrigation water	5
Optimize agronomic practices	4
Change cropping patterns, especially by altitude	4
More modern irrigation technologies	3

Table 3.6 Stakeholder-Ranked National-Level Climate Adaptations

Adaptation option	Points
Provide low interest, long-term loans to farmers	81
Create crop insurance program	71
Establish local markets	39
Improve farmer access to agronomic technology and information	34
Improve extension services	33
Improve hydrometeorological capacity	24
Produce local seeds within region	8
More direct linkage between government and farmers	4
Obtain more modern irrigation technologies	3

the need to expand farmer support services such as hydrometeorological, market access, and extension services.

In general, the stakeholder consultations revealed that farmers in Armenia have observed the changing climate and have begun responding in a variety of ways. Many have begun planting crops earlier, moving their crops to higher elevation areas, changing crop rotations, and changing the timing of irrigation for their fields. Climate change clearly challenges Armenian farmers' adaptive capacity. The combination of droughts, frost, hail, and temperature increase is especially disruptive. While the current on-farm adaptation responses have been partially successful, new programs, policies, and infrastructure investments are needed. These include crop insurance, improved hydrometeorological forecasts, improved water storage, and irrigation systems, as well as farmer training and information access about weather-related farming practices.

Evaluation and Prioritization of Adaptation Options

The menu of adaptation options to improve the resilience of Armenia's agricultural sector to climate change is derived from the results of the stakeholder consultations described in the previous section, in addition to the quantitative modeling, qualitative analysis, and expert input from international and local teams. The results reflect the following set of five criteria for prioritizing from among a larger menu of farm-level, infrastructure, programmatic, and indirect adaptation options: (1) net economic benefits (benefits minus costs); (2) qualitative expert assessment; (3) potential to aid farmers with or without climate change, referred to as "win-win" potential; (4) greenhouse gas (GHG) emissions mitigation potential; and (5) evaluation by stakeholders. Some of the options identified may also yield benefits in the form of reduced GHG mitigation potential, helping contribute to climate change mitigation as well as agricultural adaptation.

Benefit-Cost Analysis

The study conducted quantitative benefit-cost (B-C) analyses for the following eight adaptation options: (1) improving irrigation capacity and efficiency by new investments or rehabilitation to optimize application of irrigation water, (2) shifting to new crop varieties, (3) optimizing fertilizer application, (4) improving hydrometeorological services, (5) improving extension services, (6) optimizing basin-level application of irrigation water, (7) adding water storage capacity, and (8) installing hail nets for selected crops.

The results of the B-C analysis for rehabilitating irrigation infrastructure are presented in table 3.7 as an illustration of economic analyses conducted for the above options in all four agricultural regions. The table shows the B-C ratios for each crop assessed under the baseline and each climate scenario, using average price assumptions. B-C ratios above one (green shading) are favorable (that is, benefits outweigh costs), while B-C ratios below one (no shading) are not favorable (that is, costs outweigh benefits). The higher the B-C ratio (darkest green shading), the better the option is from a B-C standpoint. For example, for rainfed apricots in the intermediate agricultural region, the costs of rehabilitating irrigation infrastructure outweigh the benefits under all climate scenarios, and therefore this option is not favorable. On the other hand, for rainfed tomato in the intermediate agricultural region, the benefits of rehabilitating infrastructure far outweigh the costs under all climate scenarios, and therefore this option is favorable.

Table 3.7 Benefit-Cost Ratios for Rehabilitating Irrigation Infrastructure in Armenia's Intermediate Agricultural Region

| Irrigated/rainfed | Crop | Climate impact scenarios | | | |
		Base	Low	Medium	High
Rainfed	Alfalfa	0.60	0.60	0.60	0.60
	Apricot	0.02	0.02	0.10	0.30
	Grape	0.50	0.70	3.00	4.00
	Potato	5.00	5.00	6.00	6.00
	Tomato	21.00	23.00	27.00	27.00
	Watermelon	8.00	8.00	11.00	11.00
	Wheat	0.02	0.02	0.02	0.02

Source: World Bank data.
Note: Results are the estimated benefit-cost (B-C) ratios associated with the rehabilitation of irrigation infrastructure, by crop and climate scenario. B-C ratios greater than 1 (shaded in green) indicate that the benefits of the adaptation measure exceed the costs, while benefit-cost ratios less than 1 (not shaded) indicate that the costs exceed the benefits. Values shaded darker represent the biggest increases.

Assessment of GHG Mitigation Potential of Adaptation Options

Many of the potential adaptive measures also yield co-benefits in the form of climate change mitigation. For example, some adaptive practices can significantly reduce nitrous oxide and methane emissions. Nitrous oxide emissions are largely driven by fertilizer overuse, which increases soil nitrogen content and generates nitrous oxide. By improving fertilizer application techniques, nitrous oxide emissions can be reduced while maintaining crop yields, specifically through more efficient allocation, timing, and placement of fertilizers. Mitigation of methane emissions, on the other hand, is largely enabled by increasing the efficiency of livestock production. Optimizing breed choices, for example, serves to increase productivity, thereby reducing overall methane emissions. Alternative uses of animal manure (for example, biogas production) and improved feed quality quickens digestive processes, resulting in reduced methane emissions. Finally, adaptive measures, such as conservation agriculture and manual weeding, may also reduce the emissions associated with agricultural production and by heavy machinery use. Similarly, increased irrigation efficiency reduces the energy required to pump groundwater.

The potential for adaptive agricultural practices to simultaneously mitigate climate change has already garnered attention in Armenia. Armenia, as a transition country (United Nations Framework Convention on Climate Change [UNFCCC] Non-Annex 1, that is, not obligated by GHG emissions caps), has submitted two National Communications to the United Nations Framework Convention on Climate Change (UNFCCC 2010), and some of the Armenian Government's current agricultural policies address adaptation and mitigation priorities in the agricultural sector. Some mitigation projects in Armenia are already under way.

One World Bank project that addresses mitigation is the Natural Resources Management and Poverty Reduction Project in Armenia, which promotes the adoption of sustainable natural resource management practices and the alleviation of rural poverty in places where severe environmental degradation has occurred. The global environmental objective is to preserve the mountain, forest, and grassland ecosystems in the South Caucasus through enhanced protection and sustainable management. Specifically, to mitigate climate change, the project proposes demonstrations of biogas production installations that would reduce methane emissions while reducing the use of timber. In addition, Armenia has several projects funded through the Clean Development Mechanism, which allows Annex I countries to implement mitigation projects in non-Annex I countries (UNFCCC 2010).

National Conference

The National Dissemination and Consensus-Building Conference, held in Yerevan in October 2012, provided another opportunity to consult with Armenia's experts to identify the highest priority adaptation and mitigation options at both the national and agricultural region levels. The overall program included a detailed presentation of the technical and farmer consultation findings (outlined in the last section), and a half-day consensus-building exercise among participants, with region-focused groups providing rankings and information for the multi-criteria assessment calculations.

The small groups were presented with tables that summarized the results of the completed B-C analysis, expert assessment, win-win assessment, and mitigation assessment. The agenda for the process was in three parts: (1) rank the actions/policies for the focus region from the provided table in order of importance, including crossing off any options that are not relevant, identifying other actions or policies that should be considered, and ranking the resulting overall set of options; (2) rate the importance of three technical criteria by allocating 100 total points across: (a) B-C analysis (net economic benefit), (b) potential to help with or without climate change (win-win potential), and (c) GHG mitigation potential, to reflect the relative importance the group places on achieving each objective; and (3) report back on findings to the full conference in plenary session.

Rankings of the groups, as reported from the conference, are presented in table 3.8. The national group focused on national-scale policies, and as a result

Table 3.8 Ranking of Adaptation Measures for Armenia's Agricultural Regions

Adaptation measure	Specific focus area	Ranking of measure by group			
		National	Lowlands	Intermediate	Mountainous
Improve farmer access to agronomic technology and information	Crop varieties; more efficient use of water	1			
Create crop insurance program	Promote investments in agricultural crops susceptible to drought and hail	2			
Increase the quality, capacity, and reach of extension services	Demonstration plots	3			
Improve farmer access to hydro-meteorological capacity	Develop short-term temperature and precipitation forecasts	4			
Improve irrigation water availability	Rehabilitate irrigation capacity		1		2
Optimize agronomic practices	Increase and improve fertilizer application		4	No group formed, no ratings	
Improve crop varieties	Introduce drought-tolerant varieties		2		3
Research and improve livestock nutrition, management, and health	Include research on sheltering techniques				4
Optimize and/or improve irrigation techniques	Sprinkler, drip irrigation				2
Construct small volume reservoirs for water storage			3		5
Improve agricultural practices	Increase capacity, knowledge, and pasture management skill				1

Note: Items without entries were not ranked by those groups.

Building Resilience to Climate Change in South Caucasus Agriculture
http://dx.doi.org/10.1596/978-1-4648-0214-0

presented an entirely different focus from the region-focused groups. The region-focused groups provided additional measures for consideration unique to their regions. Across the regions, there was broad support for improving irrigation water availability, optimizing irrigation practices, and building small-scale reservoirs. No group was formed to consider the intermediate region.

Final Menu of Recommended Adaptation Options

The final menu of recommended adaptation options for Armenia reflects multiple lines of quantitative and qualitative analysis of potential net benefits, including evaluations and recommendations from farmers, stakeholders, and other experts. These measures were identified as important both at the national conference and at the farmer workshops. The six national-level measures (figure 3.5) focused on the following areas:

- *Improve farmer access to agronomic technology and information.* Through improved extension services, farmers could access technologies to improve crop yields—for example, obtaining new seed varieties or investing in drip irrigation. More targeted and practical trainings, such as demonstration plots, could lead to the use of better technologies and agronomic practices.

- *Investigate options for crop insurance, particularly for drought.* Crop insurance is not viable for the vast majority of agricultural producers due to its

Figure 3.5 National-Level Priority Adaptation Measures for Armenia

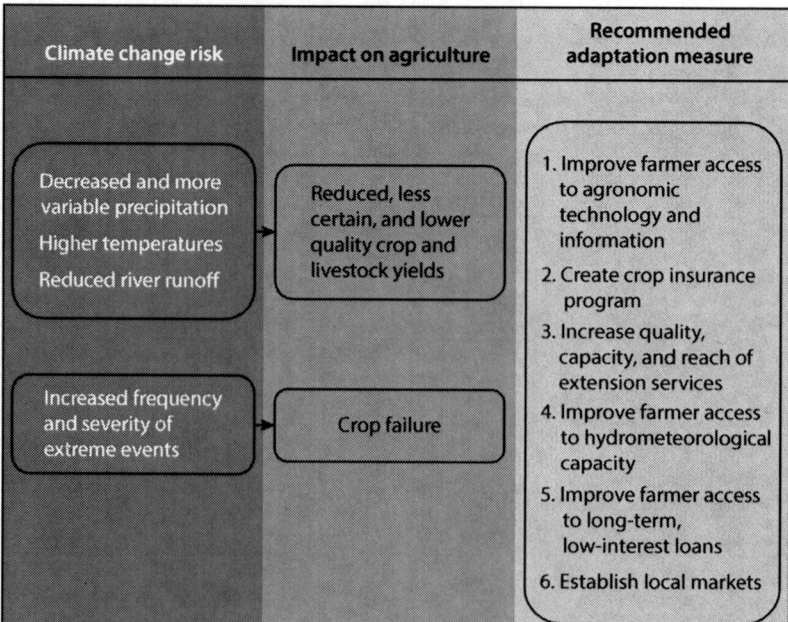

Building Resilience to Climate Change in South Caucasus Agriculture
http://dx.doi.org/10.1596/978-1-4648-0214-0

high cost, but farmers remain eager to explore insurance options. One possible way to expand coverage might be piloting a privately run weather index-based insurance program. This approach has many potential advantages over traditional multiple-peril crop insurance, including simplification of the product, standardized claim payments to farmers in a district based on the index, avoidance of individual farmer field assessment, lower administrative costs, timelier claim payments after loss, and easier accommodation of small farmers within the program. The drawback of an index-based approach may be the inability to readily insure coverage of damage from pests. In addition, pilot insurance schemes based on weather indices have encountered low demand in many locations, partly because poor farmers are cash and credit constrained; therefore they cannot afford premiums to buy insurance that pays out only after the harvest (Binswanger-Mkhize 2012). Poorly designed insurance schemes may also slow autonomous adaptation by insulating farmers from climate-induced risks. In general, countries may need to first consider improving market access and reducing credit constraints in order to better create enabling conditions suitable for crop insurance to be effective.

- *Improve the quality, capacity, and reach of the extension service, both generally and for adapting to climate change.* There was broad agreement among those surveyed that the capacity of the existing extension and research agencies must be improved to support agronomic practices at the farm level, including implementation of more widespread demonstration plots and increased access to better information on the availability and best management practices of high-yield crop varieties. The study's economic analysis suggests that expansion of extension services is very likely to yield benefits in excess of estimated costs.

- *Improve capacity of hydrometeorological institutions.* Farmers noted the need for better local capabilities for hydrometeorological data, particularly for short-term temperature and precipitation forecasts. Those capabilities are acutely needed in the short term to support better farm-level decision-making. The economic analysis of the costs and benefits of a relatively modest hydrometeorological investment, which includes training and annual operating costs, suggests that benefits of such a program are very likely to exceed costs.

- *Improve farmers' access to rural finance to enable them to access new technologies.* Farmers could acquire technologies through well-targeted and affordable credits to improve crop and livestock yields. However, the current rural finance system, with its relatively high interest rate combined with stringent collateral requirements and limited outreach, prohibits access to credit for many rural households despite the demand. The commercial banks and non-bank financial institutions (NBFI) need to tailor their loan products to the specificities of rural investments: reduce periodicity of cash-flow, provide longer maturity to match the specific crop and livestock production cycles, and pay non-monthly payments. The need for tailoring techniques to shifting climatic conditions without harming ecosystems of the country is pressing and urgent.

- *Improve access to local markets.* Specific recommendations to improve the marketability of produce and livestock in rural areas of Armenia include the following:
  - Change farmers' perception of marketing: Train them to focus on quality of products that they produce. Poor quality is not marketable, or if marketed, a low price for poor quality is inevitable.
  - Invest in market information gathering and dissemination, including mass media, fax, telephone, and real-time computer access systems.
  - Create, train, and support producer associations (cooperatives) and small and medium scale enterprises to improve the bargaining power of small farmers.
  - Provide storage facilities including cold storage that enable farmers to inventory their products for periods when the market is not saturated.

As indicated in figure 3.5, these measures address the climate change risks and corresponding impacts on agriculture. In addition, they are responsive to a key policy focus area for Armenia that was established early in the stakeholder process: Specifically, as described in box 3.1, many farms in Armenia's mountainous agricultural region operate small-scale cereal/fodder/livestock production systems, and a key policy objective for poverty relief in the country is to support these systems. Providing improved extension services and access to local markets—both measures identified above as priorities at the national level—can potentially contribute to this goal.

Box 3.1 Policy Focus Area for Armenia: Smallholder Cereal and Livestock Production

The Armenian agricultural sector is dominated by production of irrigated fruits and vegetables, particularly in the productive Ararat Valley region. A key policy objective for poverty relief, however, is support for rural subsistence farmers in the more mountainous areas, where many farms operate small-scale cereal/fodder/livestock production systems. In the early part of the 21st century livestock made up more than half of total agricultural production, but since then, crop production has grown faster than livestock production, and currently livestock is less than 40 percent of total production. Most of the crop production increases have occurred in lowland and intermediate areas, while livestock production in higher elevation areas has remained strong, with a recent focus on increases in sheep and goat production (mainly for the growing Iranian export market) (ArmStat 2013; Welton, Asatryan, and Jijelava 2013).

---

box continues next page

Box 3.1 Policy Focus Area for Armenia: Smallholder Cereal and Livestock Production *(continued)*

Climate change may be good news for farmers focusing on livestock and cereal production in high elevation areas, but only if market access can be improved. For example, crop modeling for this study found that alfalfa yields would decline by a very small amount through 2050 under the Medium Impact Scenario (1 percent rainfed, 2 percent irrigated), and wheat yields would likely increase by more than 33 percent over the same period. Although pasture was not modeled in the Armenia study, in the high elevation areas of Georgia and Azerbaijan that border Armenia, climate change was forecasted to increase pasture yields by 11 percent (western areas of Azerbaijan) to 87 percent (eastern areas of Georgia). Increases in both wheat and pasture productivity could provide a boost to smallholder cereal/livestock producers.

Farmers in these high elevation areas, however, have the greatest difficulties bringing goods to market in Armenia, not only mostly because of poor road conditions, but also because of lack of storage facilities and market knowledge, as well as the fact that export markets for landlocked Armenia have been limited in recent years. Furthermore, this study's farmer consultations in highland regions suggested that most smallholders have limited knowledge of modern livestock production techniques. Enhanced extension in these areas, coupled with greater market access, could be critical factors in unlocking the potential for higher livestock productivity in these smallholder systems.

Recommended Adaptation Options by Agricultural Region

Recommendations for each agricultural region to improve the resilience of Armenia's agricultural sector to climate change—presented in figures 3.6 through 3.8—include the following focus areas:

- *Irrigation.* All regions identified irrigation as a key focus area for improving resilience to climate changes and extremes, now and in the future. Specific measures discussed include: (1) improving existing irrigation schemes, (2) improving water use efficiency by investing in drip and sprinkler irrigation, (3) rehabilitating water reservoirs (mainly in lowland and intermediate regions), and (4) increasing national water storage capacity, in part through building small-scale reservoirs in vulnerable higher elevation regions.

- *Hydrometeorological forecasts.* Farmers currently use forecasts made available by television, but these are aimed at too broad a geographic area and do not provide information specific for agriculture (for example, information that would allow them to know when to apply pesticides, when to irrigate, or when to plant). Today, many farmers still plant when the snow is at a certain level on Mount Ararat.

- *Extension services.* The extension service run by the Armenian Government is active and well-funded, but few farmers seem to use the training or other

Figure 3.6 Lowlands Agricultural Region Priority Adaptation Measures for Armenia

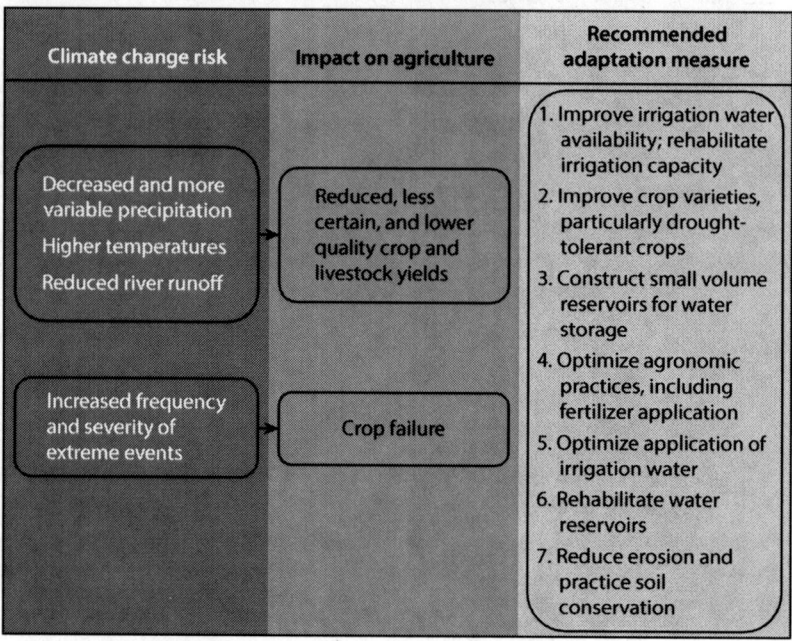

Figure 3.7 Intermediate Agricultural Region Priority Adaptation Measures for Armenia

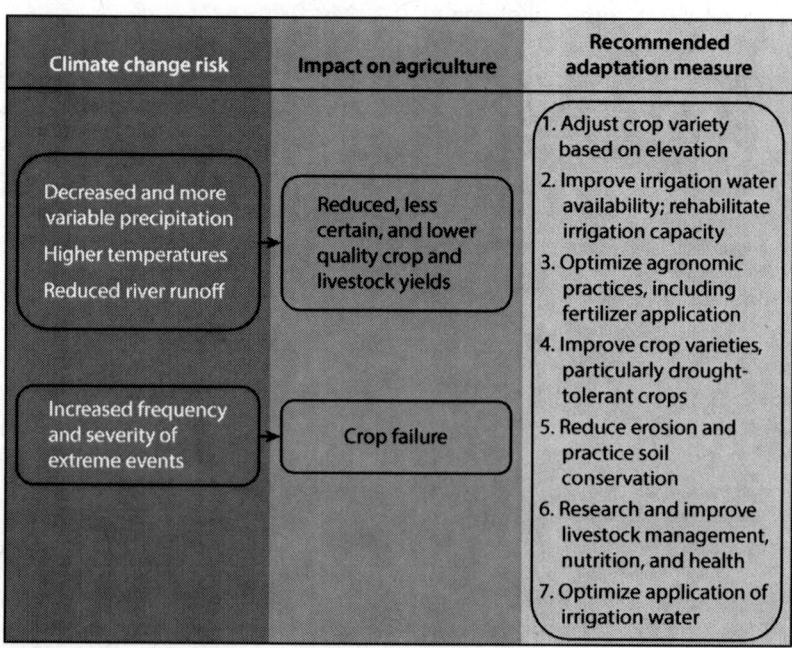

Figure 3.8 Mountainous Agricultural Region Priority Adaptation Measures for Armenia

Climate change risk	Impact on agriculture	Recommended adaptation measure
Decreased and more variable precipitation Higher temperatures Reduced river runoff	Reduced, less certain, and lower quality crop and livestock yields	1. Improve agricultural knowledge and practices 2. Improve irrigation water availability; rehabilitate irrigation capacity 3. Adopt more modern irrigation technologies 4. Improve crop varieties 5. Research and improve livestock management, nutrition, and health 6. Construct small-scale dams 7. Provide agricultural equipment
Increased frequency and severity of extreme events	Crop failure	

educational opportunities offered by the service. The farmers indicated that they would be interested in more practical and targeted training, such as demonstration plots.

- **Seed selection.** Some farmers indicated that their seedlings and plants are tolerant to weather changes, but most said they were not tolerant. Generally, farmers prefer to produce and use their own seeds, and they will clean and replant seeds from season to season. Sometimes they use seeds from the extension service, but these are often not tailored to the specific climate and soil conditions of their region. Ideally, the service would provide seeds for heat and drought tolerant crops to address anticipated warmer and drier conditions.

- **Crop insurance.** While insurance does exist, it is currently too expensive for most farmers. Both hail and spring frost are major issues for farmers in the region, with estimates of annual losses on the order of 10 percent of annual production for some crops, which may account for as much as US$100–150 million in annual losses nationwide. Subsidized programs for crop insurance would greatly stabilize their incomes and improve their capacity to reinvest in farming, but insurance schemes must be carefully designed for affordability, and they must recognize cash and credit constraints if there is to be sufficient uptake of insurance among poor smallholder farmers.

- **Bank loans.** Most farmers indicated they have access to high-interest, short-term bank loans for agricultural development, but it is difficult to obtain low-interest, long-term bank loans for agricultural development.

- **Infrastructure.** To moderate temperatures and improve yields, some farmers have constructed greenhouses. Few farmers attending the stakeholder meeting had greenhouses, however, as most of these farmers were smallholders.

Limitations of the Study

Finally, due to its broad scope, this study necessarily involves significant limitations. These include the need to make simplifying assumptions about many important aspects of agricultural and livestock production in Armenia, and the limitations of simulation modeling techniques for forecasting crop yields and water resources. As a result, certain recommendations may require a more detailed examination and analysis than could be accomplished here in order to ensure that specific adaptation measures are implemented in a manner that maximizes their value to Armenian agriculture. However, the authors hope that the awareness of climate risks and the analytic capacities built over the course of this study provide not only a greater understanding among Armenian agricultural institutions of the basis of the recommendations presented here, but also an enhanced capability to conduct the required more detailed assessment that will be needed to further pursue the recommended actions.

In addition, it is desirable that the countries of the South Caucasus address climate change through collaboration on issues such as climate-related data sharing and crisis response. There are many challenges to achieving these objectives, but fortunately there are a wide range of existing models of regional-scale institutional arrangements throughout the world, encompassing the scope of regional cooperation for water resources planning, agricultural research and extension, and enhanced hydrometeorological service development and data provision.

References

ArmStat (Republic of Armenia, National Statistical Service). 2013. Various regional statistics publications, as listed on the website (accessed December 9, 2013), http://www.armstat.am/en/?nid=50.

Binswanger-Mkhize, H. P. 2012. "Is There Too Much Hype about Index-Based Agricultural Insurance?" *Journal of Development Studies* 48 (2): 187–200.

Stokes, C. R., S. D. Gurney, M. Shahgedanova, and V. Popovnin. 2006. "Late-20th-Century Changes in Glacier Extent in the Caucasus Mountains, Russia/Georgia." *Journal of Glaciology* 52 (176): 99–109.

Thornton, P. K., J. van de Steeg, A. Notenbaert, and M. Herrero. 2009. "The Impacts of Climate Change on Livestock and Livestock Systems in Developing Countries: A Review of What We Know and What We Need to Know." *Agricultural Systems* 101: 113–27.

UNFCCC (United Nations Framework Convention on Climate Change). 2010. *Second National Communication on Climate Change: A Report under the United Nations Framework Convention on Climate Change*. Republic of Armenia, Ministry of Nature Protection, Yerevan.

Welton, G., A. Asatryan, and D. Jijelava. 2013. *Comparative Analysis of Agriculture in the South Caucasus*. United Nations Development Programme, Tbilisi, Georgia.

WWF (World Wildlife Fund Norway, and WWF Caucasus Programme). 2009. "Climate Change in Southern Caucasus: Impacts on Nature, People and Society." Report, WWF Norway, Oslo (accessed October 7, 2013), http://assets.wwf.no/downloads/climate _changes_caucasus___wwf_2008___final_april_2009.pdf.

CHAPTER 4

Azerbaijan: Risks, Impacts, and Adaptation Menu

This chapter summarizes the results of efforts to develop a menu of adaptation options for the agricultural sector in Azerbaijan. It is organized into four sections: (1) climate risk, (2) climate impacts, (3) adaptation assessment, and (4) evaluation and prioritization of adaptation options.

Climate Risk

Historical Climate Trends

The South Caucasus region has seen a variety of changes in climate, including increasing temperatures, shrinking glaciers, sea level rise, reduction and redistribution of river flows, decreasing snowfall, and an upward shift of the snowline. In the past 10 years, the region has also experienced more extreme weather events—flooding, landslides, forest fires, and coastal erosion—which have resulted in economic losses and human casualties (WWF 2009).

Figures 4.1 and 4.2 present historical temperature and precipitation data for Azerbaijan. Figure 4.1 shows annual temperatures and growing season temperatures, 1900–2012. During 1980–2012, average annual temperature increased 1.2°C, while average growing season temperature increased 1.3°C. Figure 4.2 presents average monthly precipitation and average growing season precipitation, 1900–2012.

The increasing temperatures have caused the glaciers to melt rapidly in the region, as has been occurring globally. The volume of glaciers in the South Caucasus has been reduced by 50 percent over the last century, and 94 percent of the glaciers retreated 38 meters per year (Stokes et al. 2006). In Azerbaijan the main glacier areas are in Gusarchay Basin in the Greater Caucasus. The area of glaciers has decreased from 4.9 to 2.4 square kilometers (km^2) in the past 110 years. Natural water resources are declining, and therefore, water shortages are becoming more frequent.

Figure 4.1 Average Annual and Growing Season Temperatures in Azerbaijan, 1900–2012

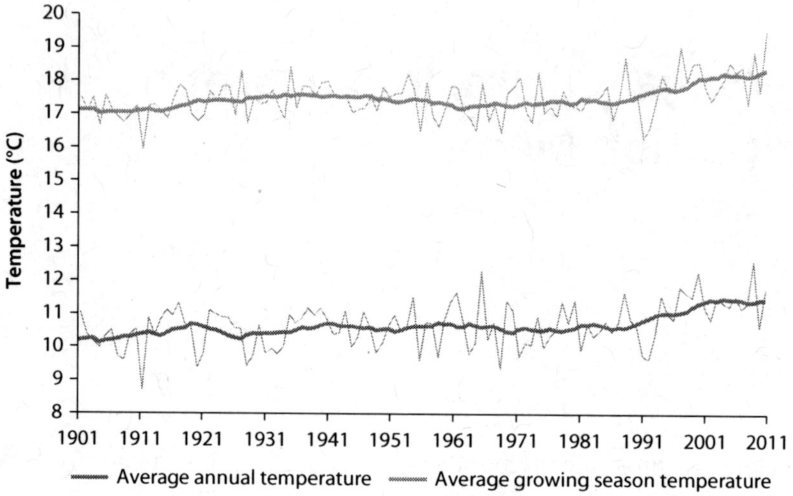

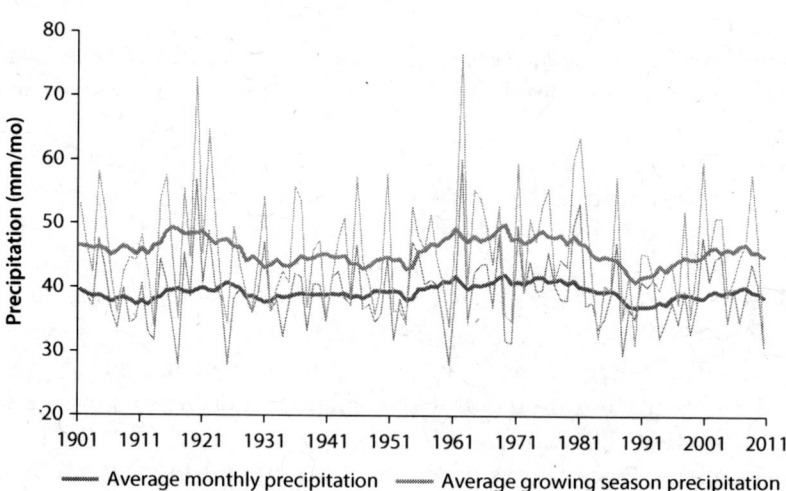

Source: University of East Anglia Climatic Research Unit, Norwich, UK.

Figure 4.2 Average Monthly and Growing Season Precipitation in Azerbaijan, 1900–2012

Source: University of East Anglia Climatic Research Unit, Norwich, UK.
Note: mm/mo = millimeters per month.

Forecasted Changes in Temperature and Precipitation

This study's results reveal a gradual increasing trend in temperature that will accelerate in the near future for the four agricultural regions of Azerbaijan—high rainfall, low rainfall, irrigated, and subtropical— (map 4.1), and farmers have also observed an increase in the frequency of extreme heat events. This trend is

Map 4.1 Predicted Effect of Climate Change on Average Annual Temperature in the 2040s

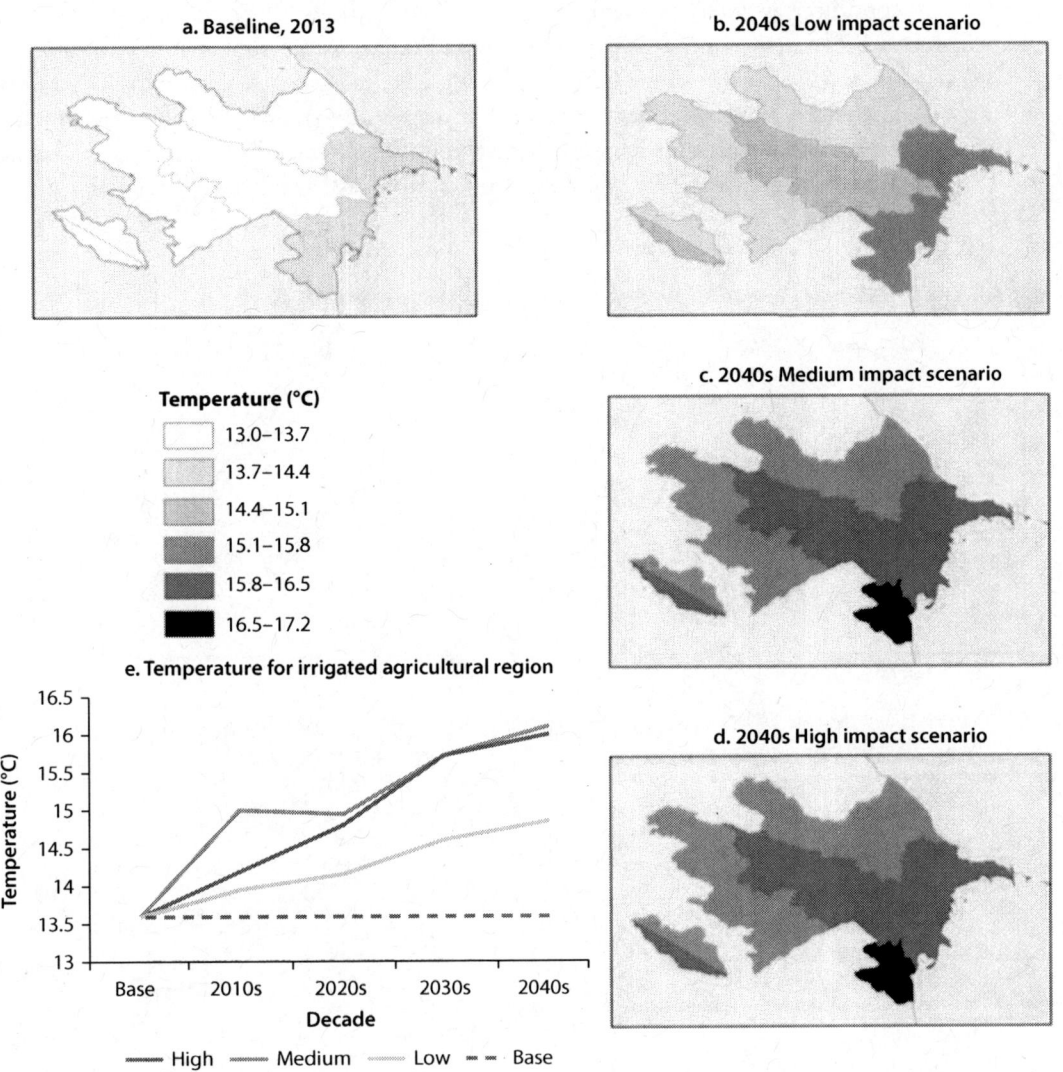

a. Baseline, 2013

b. 2040s Low impact scenario

c. 2040s Medium impact scenario

d. 2040s High impact scenario

Temperature (°C)

- 13.0–13.7
- 13.7–14.4
- 14.4–15.1
- 15.1–15.8
- 15.8–16.5
- 16.5–17.2

e. Temperature for irrigated agricultural region

High   Medium   Low   Base

Sources: ©Industrial Economics. Used with permission; reuse allowed via Creative Commons Attribution 3.0 Unported license (CC BY 3.0). Country boundaries are from ESRI and used via CC BY 3.0.

consistent with the observed historical trend and information gathered from local farmer workshops. The average increase in temperature over the next 50 years, is estimated to be about 2.4°C, compared with the 0.75°C increase in temperature observed in the western portion of Azerbaijan from 1961 to 2000 (UNFCCC 2010). Though the degree of warming that will occur in the country remains uncertain, and warming could be more modest than 2.4°C, even the Low Impact Scenario predicts an average temperature increase of 1.3°C, compared with current conditions. The differences in temperature among Azerbaijan's four agricultural regions are small, and the warming trend relative to current conditions is about the same magnitude across the four agricultural regions.

Estimates of how precipitation will change in Azerbaijan are much more uncertain than those for temperature. By 2050 all scenarios indicate uncertainty in the direction of effect as well as its magnitude (map 4.2). The Low Impact Scenario forecasts an increase in precipitation, while the other two scenarios predict decreases. The Medium Impact forecast predicts a national decline

Map 4.2 Effect of Climate Change on Average Annual Precipitation in the 2040s

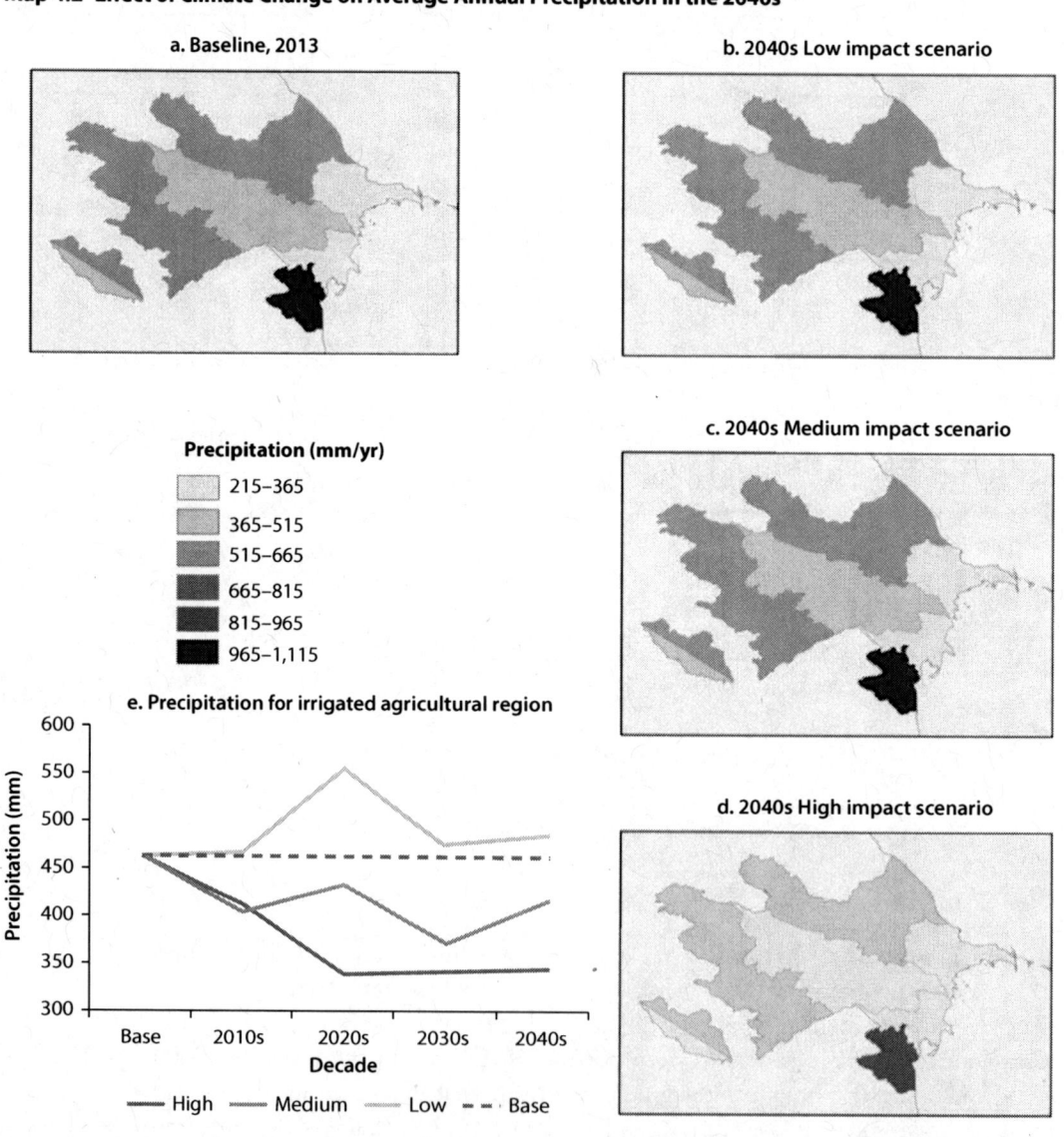

Sources: ©Industrial Economics. Used with permission; reuse allowed via Creative Commons Attribution 3.0 Unported license (CC BY 3.0). Country boundaries are from ESRI and used via CC BY 3.0.
Note: mm = millimeters.

in precipitation of about 41 millimeters (mm) per year, most occurring in the high rainfall agricultural region, while the High Impact Scenario predicts an almost 20 percent decline in precipitation by 2050. Uncertainty at the regional level is even higher, and annual precipitation declines in the subtropical agricultural region could be as large as 160 mm per year.

Inundations and flash floods are already common in Azerbaijan's high water season, and the frequency of these extreme events has been increasing in the past two decades. Climate change is likely to increase the frequency and magnitude of flooding. Even though precipitation is expected to increase only in the Low Impact Scenario by the 2040s (map 4.2), rainfall events are predicted to be more variable, with a high probability of daylong to multi-day events being larger and less frequent. An increase in flooding could pose challenges to Azerbaijan's agriculture sector, particularly in the spring when flooding can delay or prevent planting of summer crops, and during the late summer, when flooding can destroy the entire year's growth and prevent timely harvesting. Smaller flood events can also reduce crop productivity through waterlogging. And regardless of when they take place, floods can cause loss of top soil and agricultural land and contribute to erosion.

The seasonal distribution of temperature and precipitation are more important for agricultural production than the overall yearly averages for these variables. Predicted temperature increases for Azerbaijan are greatest during August–October relative to current conditions, and the potential summer temperature increase is predicted to reach as high as 4°C in the subtropical agricultural region of the country, when temperatures are already highest. Forecasted precipitation declines are also estimated to be greatest in the key April–October period. Figure 4.3 presents the monthly baseline and forecasted temperatures and precipitation for each scenario for the irrigated agricultural region.

Climate Impacts

In order to assess the impact of climate change on the agricultural sector in Azerbaijan, the monthly projections of temperature and precipitation were translated to daily projections for use in crop models, as described in chapter 2, box 2.2. The crop models examined the potential effect of climate change on crop yields in Azerbaijan under the "no adaptation" scenario (that is, if no adaptation measures are taken). The crop yield impacts presented in table 4.1 represent the potential outcome under the Medium Impact Scenario and do not take into account irrigation water constraints.

Decline in Crop Yields

As shown in table 4.1, yields of all key crops in Azerbaijan's agricultural sector (aside from pasture) will generally decrease across agricultural regions as a result of rising temperatures and water stress. Rainfed potato and cotton are expected to experience the greatest declines in yield. Pasture yields, on the other hand, are predicted to significantly increase in all four agricultural regions, particularly in the high rainfall and subtropical agricultural regions.

Figure 4.3 Predicted Effect of Climate Change on Monthly Temperature and Precipitation Patterns for the Irrigated Agricultural Region, 2040s

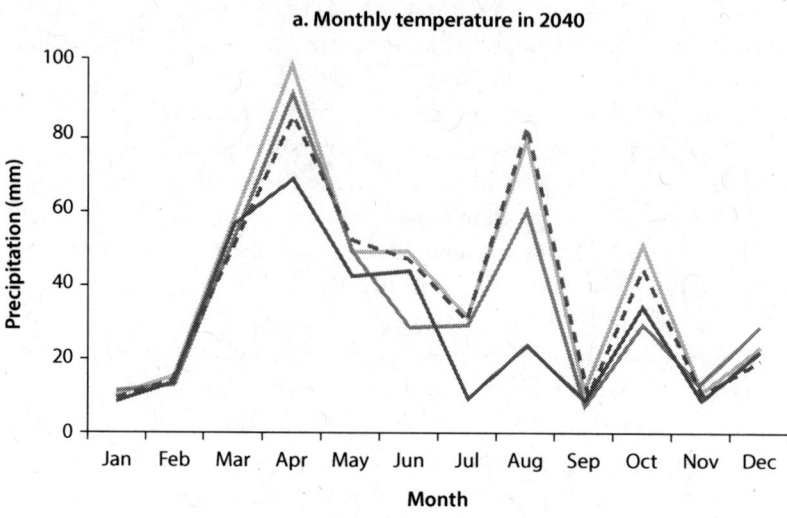

a. Monthly temperature in 2040

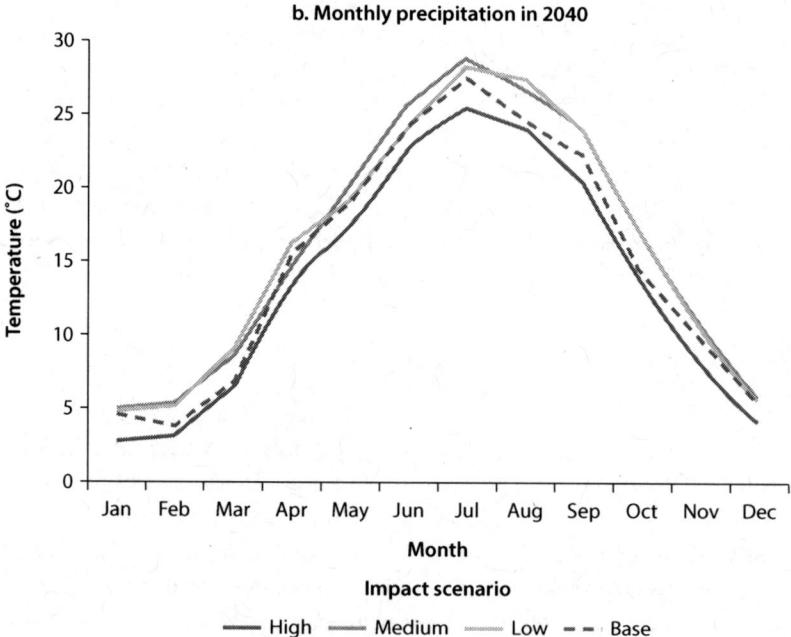

b. Monthly precipitation in 2040

Impact scenario

—— High   —— Medium   ⋯⋯ Low   – – Base

Source: World Bank data.
Note: mm = millimeters.

Although table 4.1 reflects the assumption that irrigation water will not be constrained, changes in temperature and precipitation resulting from climate change are expected to impact water resources in Azerbaijan. As a result, a more detailed water resource analysis is also needed to determine the extent of climate change impacts. The study team conducted a water availability analysis for Azerbaijan using the Water Evaluation and Planning System (WEAP) and the

Table 4.1 Effect of Climate Change on Crop Yields in the 2040s under the Medium Impact Scenario, No Adaptation and No Irrigation Water Constraints

Irrigated/rainfed	Crop	Change in yield (%)			
		High rainfall	Irrigated	Low rainfall	Subtropical
Irrigated	Alfalfa	–7	–7	–6	–2
	Corn	–6	–7	–6	–6
	Cotton	–1	–3	–4	–5
	Grape	–5	–5	–5	–5
	Potato	–7	–9	–5	–6
	Wheat	–5	–5	–5	–5
Rainfed	Alfalfa	–6	–8	–6	–8
	Corn	2	–7	–7	–6
	Cotton	–13	–13	–13	–10
	Grape	–7	–16	–5	–6
	Pasture	11	5	6	11
	Potato	–12	–13	–14	–11
	Wheat	–5	–6	–5	–5

Source: World Bank data.
Note: Results are average changes in crop yield, assuming no effect of carbon dioxide fertilization. Declines in yield are shown in shades of orange, with darkest representing biggest declines; increases are shaded green, with darkest representing the biggest increases.

Climate and Runoff Hydrologic Model (CLIRUN) model. Next, water supply estimates were matched with forecasts of water demand for all sectors, including agriculture, to determine water availability. Agricultural water demand was estimated using the AquaCrop model (see chapter 2, box 2.2 for more information).

Water Supply Declines, Demand Increases

Figure 4.4 presents the estimate effect of climate change on mean monthly runoff in Azerbaijan in the 2040s. The runoff indicator is directly relevant to agricultural systems and provides insight into the risk of climate change for agricultural water availability, as well as the implications of climate change for water resource management. As shown in figure 4.4, relative to current estimates, runoff is predicted to decline under the High and Medium Impact Scenarios after 2030, but it is expected to increase under the Low Impact Scenario. Variability across the scenarios increases significantly after 2020. In terms of monthly effects, though annual runoff under the Low Impact Scenario is forecasted to increase, runoff during the late spring and late summer months declines under all three scenarios relative to baseline conditions. Agricultural demand for water is already at its highest during these months, and furthermore, AquaCrop forecasts an increase in demand for water under climate change during this time of the year as well.

The results indicated that irrigation water shortages already occur under the baseline and rise significantly under climate change. In all scenarios, over 67 percent of irrigation demands were unmet in the Lenkeran/Southern Caspian and Eastern Lower Kur basins by the 2040s. The FAO (Food and Agriculture Organization of the United Nations) crop sensitivity factors were used to estimate

Figure 4.4 Estimated Climate Change Effect on Mean Monthly Runoff in the 2040s for All Azerbaijani Basins

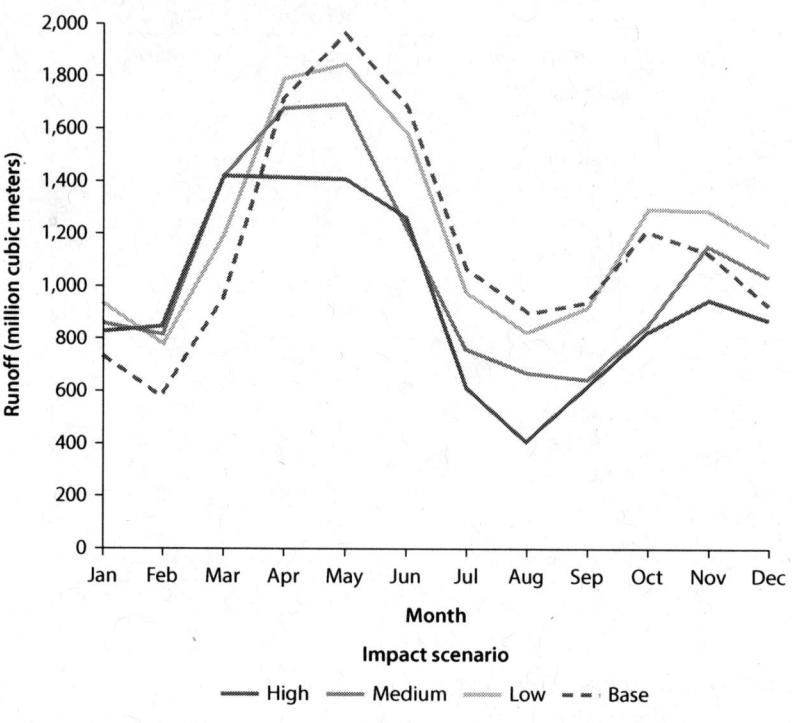

Source: World Bank data.

the change in yield resulting from a reduction in water availability for each crop, agricultural region-basin area, and climate scenario. The significant reductions in irrigation water availability due to climate change led to large predicted declines in the yields of irrigated crops of up to –77 percent (see table 4.2).

Negative Net Climate Effects

Therefore three climate change stressors combine to yield an overall negative impact on crop yields in Azerbaijan:

- The direct effect of temperature and precipitation changes on crops
- Increased irrigation demand required to maintain yields
- Decline in water supply associated with higher evaporation and lower rainfall

All of these effects have a more pronounced impact during the summer growing season. The net effect of these three factors on irrigated agriculture in Azerbaijan is illustrated in table 4.2. Panel a of the table shows the effect of temperature and precipitation changes alone on irrigated agriculture if there are no irrigation water constraints. Panel b show the combined effect of all three factors mentioned above, including the forecasted irrigation water shortages. The combined, net effect of these factors on crop yields is dramatic and provides an important focus for adaptation efforts to mitigate potential losses.

Table 4.2 Effect of Climate Change on Irrigated Crop Yields Adjusted for Estimated Irrigation Water Deficits in the 2040s

a. Crop yield impacts due to temperature and precipitation changes, without considering irrigation water constraints

Crop	Change in yield (%)			
	High rainfall	Irrigated	Low rainfall	Subtropical
Alfalfa	−7	−7	−6	−2
Corn	−6	−7	−6	−6
Cotton	−1	−3	−4	−5
Grape	−5	−5	−5	−5
Potato	−7	−9	−5	−6
Wheat	−5	−5	−5	−5

b. Crop yield impacts due to temperature and precipitation changes, as well as forecasted irrigation water constraints

Crop	Change in yield (%)		
	HR	Irr	LR
Samur/N.Caspian			
Alfalfa	−27	−28	−26
Corn	−27	−27	−27
Cotton	−19	−21	−22
Grape	−23	−23	−23
Potato	−27	−29	−26
Wheat	−26	−26	−26
Ganikh			
Alfalfa	−28		
Corn	−28		
Cotton	−20		
Grape	−24		
Potato	−28		
Wheat	−27		
E.Lower Kur			
Alfalfa	−42	−43	−42
Corn	−42	−43	−42
Cotton	−33	−34	−35
Grape	−36	−36	−36
Potato	−43	−44	−42
Wheat	−42	−42	−42

(table continues next page)

Table 4.2 Effect of Climate Change on Irrigated Crop Yields Adjusted for Estimated Irrigation Water Deficits in the 2040s *(continued)*

b. Crop yield impacts due to temperature and precipitation changes, as well as forecasted irrigation water constraints

Crop	Change in yield (%)		
	Irr	LR	ST
b (4) Lenkeran/S.Capian			
Alfalfa	−77	−77	−76
Corn	−77	−77	−77
Cotton	−65	−65	−66
Grape	−66	−66	−66
Potato	−77	−77	−77
Wheat	−77	−77	−77

Source: World Bank data.
Note: Results are average changes in crop yield, assuming no effect of carbon dioxide fertilization. Declines in yield are shown in shades of orange, with darkest representing biggest declines; increases are shaded green, with darkest representing the biggest increases. HR = high rainfall; Irr = irrigated; LR = low rainfall; ST = subtropical.

By examining the potential effects of irrigation water constraints on particular crops, the study team was able to respond to key agricultural policy focus areas in Azerbaijan. For example, as described in box 4.1, stakeholders were particularly interested in the potential effects of climate change on cotton production, a crop that traditionally contributes significantly to the country's agricultural sector. As indicated in table 4.2, cotton yields in the irrigated areas of the Eastern Lower Kur basin (where more than half of existing cotton production occurs) are expected to decrease substantially in the 2040s under the Medium Impact Scenario when taking into account forecasted irrigation water shortages. This insight helped communicate the urgency of the situation to farmers and policy makers, particularly with respect to the potential impact of climate change on availability of irrigation water.

The direct effects of climate change on livestock could also be severe, but due to lack of location-specific data, this analysis does not quantify these impacts. However there is a robust literature establishing that higher temperature decreases livestock productivity. The indirect effect of climate change on livestock feed stocks, including pasture, would, according to the analysis in this study, be positive and provides a counterbalance to the negative direct heat stress effects.

Adaptation Assessment

After examining the local climate risk and likely impacts of climate change on Azerbaijan's agricultural sector, the study team conducted an adaptation assessment of the sector, both at the national and regional levels. This involved stakeholder outreach to elicit information about current farming practices, observed impacts of climate change thus far, and how farmers are currently adapting to these impacts. In addition, the stakeholder outreach sessions allowed the study

---

Box 4.1 Policy Focus Area for Azerbaijan: The Future of Cotton Production

Early stakeholder consultations with Azerbaijani counterparts revealed an interest in having the study explore the productivity of cotton farming under changed climatic conditions. In former times, cotton was a major component of Azerbaijan's agricultural sector. As recently as the 1980s, the cropped area of cotton was more than 300,000 hectare (ha). However, by 2012 less than 30,000 ha were cultivated with cotton. One reason for the decline is relatively low productivity. In the mid-1980s Azerbaijan's cotton fields produced an average of over 3 tons/ha, while productivity in 2012 was less than 2 tons/ha, and averaged closer to 1.35 tons/ha in the preceding 6 years (2006–11) (AZStat 2013). As a sign of cotton's past role in the agricultural economy, and continuing if diminished role in generating foreign currency as an export crop, the Azerbaijan National Academy of Sciences maintains a Scientific Research Institute of Cotton-Growing in Ganja. Recent trends notwithstanding, the wide access to irrigation and the potential for possibly enhanced climatic conditions have raised interest as to whether cotton production might increase as a result of climate change, and so this idea was analyzed with the crop and water resource modeling tools.

Crop modeling shows that irrigated cotton would experience modest declines in yield (1–5 percent) under the climate change Medium Impact Scenario by mid-century, provided sufficient irrigation water was available. More than half of the existing cotton production appears to be located in the valleys of the Eastern Lower Kur basin, in areas located in the irrigated agricultural region. The water resource modeling for this study showed the potential for a severe shortage of irrigation water in this area, as well as in many parts of Eastern Azerbaijan. With shortages for irrigation water in the range of 65 percent or more, it seems clear that cotton production will be threatened by climate change in those regions where it is currently grown. Furthermore, recent statistics from AZStat show that much of the cotton production in recent years has been unprofitable, with production costs exceeding revenue (AZStat 2013). One reason may be continued downward pressure on world cotton prices by Chinese policies to stockpile surplus production (Wexler 2013). In this situation, even aggressive climate adaptation is not likely to be sufficient to enhance the fortunes of most Azerbaijani cotton farmers.

---

team to compile an initial list of priority adaptation options based on input from farmers as well as government officials and other local experts. This section describes the findings of the adaptation assessment and the recommended adaptation options from the stakeholder consultations.

Current Regional Adaptive Capacity

To assess Azerbaijan's current regional adaptive capacity, it was essential that the study team inform and consult with a variety of local stakeholders—farmers and farmers' associations, local government officials, and other local experts—on the predicted impacts of climate change on agriculture and water resources. The team first met with farmers for a one day stakeholder workshop in Shamakhi in March 2012. A second set of farmer consultations were conducted in October

Map 4.3 Locations of Stakeholder Consultations in Azerbaijan

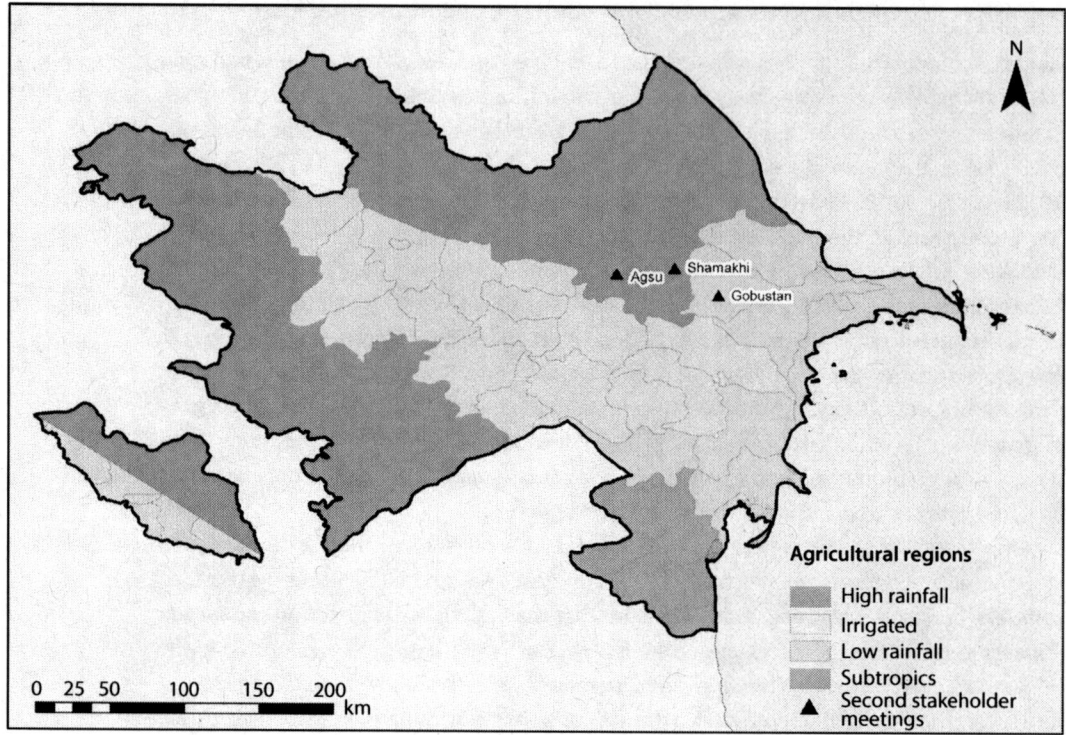

2012 at three locations (Shamakhi, Agsu, and Gobustan), representing different agricultural regions of Azerbaijan (map 4.3).

At the initial workshop, participants were given an overview of the study and the potential impacts of climate change on crop yields and water availability in Azerbaijan. They were then asked if they had witnessed such impacts and what they have done, or would do, to mitigate their effects. The stakeholders at the workshop made it clear that, although farmers are becoming more flexible in their response to climate events, their adaptive capacity is still quite limited due to poorly maintained irrigation and drainage systems, limited financial resources, and inadequate support from and access to extension services.

At the subsequent farmer consultations, participants were provided with a list of potential climate adaptations. They were asked to remove any irrelevant adaptations and to add any additional adaptations they believed would be effective.

Participants then provided rankings for both national-level and regional-level adaptation options. Adaptation rankings varied among regions, reflecting differences in their current climates, topography, and other location-specific factors. The ranked recommendations of adaptation options for Azerbaijan's high rainfall, irrigated, and low rainfall agricultural regions are as follows (consultations were not held in the subtropical agricultural region).

High Rainfall Agricultural Region: Shamakh

This region produces a variety of crops, including wheat, barley, other grains, grape, and orchard fruits, as well as livestock. The most important weather-related impact noted in this region is drought, which can be severe during the summer. Hail also affects crop production in this region, and flooding can destroy harvests.

The most important adaptation option according to farmers here is increasing and improving the application of fertilizer, although improving livestock management and crop varieties were also highly ranked alternatives (table 4.3). Rehabilitation of irrigation systems and water reservoirs were also key concerns.

Irrigated Agricultural Region: Agsu

Key crops grown in this region include winter wheat, alfalfa, fodder crops, grape, and orchard fruits such as pomegranate. Livestock are also raised in the region, mostly sheep and cattle. Most of the croplands are flood-irrigated from earth canals using water from the Kur River without pumping. Farmers indicated they receive most of the water they need, but sometimes require storing water in winter and spring for needs in the late summer and early fall. Groundwater is used in the region (20 active wells), although the bulk of water use is still sourced from surface water.

The largest issue in the region is droughts, although there are also occasional landslides and floods. Maximum temperatures rise to 40–42°C, which can cause wilting in crops. Land degradation including salinization has also been a problem in the region.

Higher ranked adaptation options (table 4.4) include increasing and improving the application of fertilizer and improving existing crop varieties for drought tolerance. Also important is the need to rehabilitate aging irrigation infrastructure and to improve the timing of irrigation water application—the latter requiring better connection to hydrometeorological forecasts as well as enhanced know-how.

Table 4.3 Ranked Recommendations from the Shamakhi Consultation

Adaptation option	Points
Increase and improve application of fertilizer	18
Improve livestock nutrition and shelter	12
Improve crop varieties, particularly those tolerant to droughts	12
Rehabilitate irrigation and drainage	7
Rehabilitate water reservoirs	4
Restore pastures by using improved grazing practices	3
Create soil maps to improve precision of fertilizer application	2
Improve pest management	2
Use hail rockets	1

Building Resilience to Climate Change in South Caucasus Agriculture
http://dx.doi.org/10.1596/978-1-4648-0214-0

Table 4.4 Ranked Recommendations from the Agsu Consultation

Adaptation option	Points
Increase and improve application of fertilizer	18
Improve crop varieties, particularly those tolerant to droughts	12
Rehabilitation of irrigation	7
Improve livestock nutrition and shelter	6
Optimize use of irrigation water	5
Rehabilitation of drainage systems	2

Low Rainfall Agricultural Region: Gobustan

Major crops in this region include wheat, barley, grape, vegetables, and pome-granate. Although most of the farming is rainfed, farmers will irrigate if water is available. Extreme events described by the participants include landslides, droughts, and wind erosion. Flooding is an occasional concern, as are heat waves in the southern part of Gobustan. According to farmers, frost events occur about once per decade. Farmers have observed that flood and drought events have gotten worse over the past several decades and that temperatures have risen.

The high rankings given to irrigation-related adaptations (table 4.5) clearly reflect the importance of irrigation to agricultural production in this region. Improved crop varieties and improved application of fertilizer were also recommended—the latter likely to require enhancement of farmers' know-how on timing and application rates. Farmers keep livestock, but have limited pasture to support them and are aware of the need to improve basic animal husbandry practices.

Current National-Level Adaptive Capacity and Responses

Participants in all three regions generally agreed about the need to improve hydrometeorological forecasting capacity. This adaptation, along with improving farmer access to extension services, were the highest ranked items of the adaptations recommended by farmers (table 4.6).

The need to expand farmer support services to crop insurance and the improvement of access to low-interest, long-term loans forms a second tier of needed enhancements. Currently available loans are often short-term and bear high interest rates. While farmers said that crop insurance was sometimes available on the private market, they could not afford to pay the premiums. They were very interested in obtaining insurance against hail and frost.

In general, the stakeholder consultations revealed that farmers in Azerbaijan have observed the changing climate and have already begun responding in a variety of ways. Many have begun planting crops earlier, moving their crops to higher elevation areas, changing crop rotations, and changing the timing of irrigation on their fields.

Climate change clearly challenges Azerbaijani farmers' adaptive capacity. While the current on-farm adaptation responses have been partially successful, new programs, policies, and infrastructure investments are needed. These include

Table 4.5 Ranked Recommendations from the Gobustan Consultation

Adaptation option	Points
Rehabilitate irrigation	18
Optimize use of irrigation water	15
Improve crop varieties, particularly those tolerant to droughts	12
Increase and improve application of fertilizer	9
Improve livestock nutrition and shelter	6
Rehabilitate drainage systems	3

Table 4.6 Stakeholder-Ranked National-Level Climate Adaptations

Adaptation option	Points
Improve hydrometeorological capacity	34
Improve farmer access to agricultural technology	27
Improve extension services	26
Create crop insurance program	18
Improve access to long-term, low-interest loans	16

crop insurance, improved hydrometeorological forecasts, improved water storage, irrigation systems, as well as farmer training and information access about weather-related farming practices.

Evaluation and Prioritization of Adaptation Options

The menu of adaptation options to improve the resilience of Azerbaijan's agricultural sector to climate change is derived from the results of stakeholder consultations described in the previous section, in addition to the quantitative benefit-cost (B-C) modeling, qualitative analysis, and expert input from international and local teams. The results reflect the following set of five criteria for prioritizing from among a larger menu of farm-level, infrastructure, programmatic, and indirect adaptation options: (1) net economic benefits (benefits minus costs); (2) qualitative expert assessment; (3) potential to aid farmers with or without climate change, referred to as "win-win" potential; (4) greenhouse gas (GHG) emissions mitigation potential; and (5) evaluation by stakeholders. Some of the options identified may also yield benefits in the form of reduced GHG mitigation potential, helping contribute to climate change mitigation as well as agricultural adaptation.

Benefit-Cost Analysis

The study team conducted quantitative B-C analyses for the following eight adaptation options: (1) improving irrigation capacity and efficiency by new investments or rehabilitation to optimize application of irrigation water; (2) shifting to new crop varieties; (3) optimizing fertilizer application; (4) improving hydrometeorological services; (5) improving extension services; (6) optimizing

Table 4.7 Benefit-Cost Ratios for Optimizing Crop Varieties in Azerbaijan's Irrigated Agricultural Region

Irrigated/rainfed	Crop	Climate impact scenarios			
		Base	Low	Medium	High
Irrigated	Alfalfa	0.1	0.1	0.1	0.2
	Corn	7.0	7.0	7.0	7.0
	Cotton	29.0	28.0	30.0	30.0
	Grape	30.0	29.0	29.0	29.0
	Pasture	0.0	0.0	0.0	0.0
	Potato	54.0	52.0	51.0	50.0
	Wheat	10.0	10.0	10.0	10.0
Rainfed	Alfalfa	0.1	0.1	0.1	0.1
	Corn	7.0	7.0	7.0	7.0
	Cotton	24.0	23.0	23.0	21.0
	Grape	29.0	29.0	27.0	26.0
	Pasture	0.0	0.0	0.0	0.0
	Potato	45.0	44.0	42.0	40.0
	Wheat	10.0	10.0	10.0	10.0

Source: World Bank data.

Note: Results are the estimated benefit-cost (B-C) ratios associated with the rehabilitation of irrigation infrastructure, by crop and climate scenario. B-C ratios greater than 1 (shaded in green, with darkest representing the biggest increases) indicate that the benefits of the adaptation measure exceed the costs, while benefit-cost ratios less than 1 (not shaded) indicate that the costs exceed the benefits.

basin-level application of irrigation water; (7) adding water storage capacity; and (8) installing hail nets for selected crops.

The results of the benefit-cost analysis for optimizing crop varieties are presented in table 4.7 as an illustration of economic analyses conducted for the above options in all four agricultural regions. B-C ratios above 1 (green shading) are favorable (that is, benefits outweigh costs), while B-C ratios below 1 (no shading) are not favorable (that is, costs outweigh benefits). The higher the B-C ratio (darkest green shading), the better the option is from a B-C standpoint. For example, for alfalfa and pasture in the irrigated agricultural region, the costs of optimizing crop varieties outweigh the benefits under all climate scenarios, and therefore this option is not favorable. On the other hand, for irrigated potato in the irrigated agricultural region, the benefits of optimizing crop varieties far outweigh the costs under all climate scenarios, and therefore this option is favorable.

Assessment of GHG Mitigation Potential of Adaptation Options

Many of the adaptive measures recommended also yield co-benefits in climate change mitigation. This section discusses the study's assessment of each option's potential for GHG mitigation and highlights the specific adaptive measures that demonstrate the greatest opportunities for emissions reductions. Adaptive practices can significantly reduce nitrous oxide and methane emissions. Nitrous oxide emissions are largely driven by fertilizer overuse which increases soil nitrogen content and generates nitrous oxide. By improving fertilizer application techniques, nitrous oxide emissions can be reduced while maintaining crop yields, specifically through more efficient allocation, timing, and placement of fertilizers.

Mitigation of methane emissions, on the other hand, is largely enabled by increasing the efficiency of livestock production. Optimizing breed choices, for example, serves to increase productivity thereby reducing overall methane emissions. Alternative uses of animal manure (for example, biogas production) and improved feed quality quickens digestive processes, resulting in reduced methane emissions. Finally, adaptive measures such as conservation agriculture and manual weeding may also reduce the emissions associated with agricultural production and by heavy machinery use. Similarly, increased irrigation efficiency reduces energy required to pump groundwater.

The potential for adaptive agricultural practices to simultaneously mitigate climate change has already garnered attention in Azerbaijan. Azerbaijan, as a transition country (UNFCCC Non-Annex 1, that is, not obligated by GHG emissions caps), has submitted two National Communications to the United Nations Framework Convention on Climate Change (UNFCCC 2010), and some of the Azerbaijan Government's agricultural policies address adaptation and mitigation priorities in the agricultural sector. Examples of legislation and actions taken by the Azerbaijan Government relevant to GHG mitigation include: installation of pilot projects of biogas facilities in four regions to raise public awareness; the National Programme on the Rehabilitation and Expansion of Forests of 2003 for reforestation and afforestation; and the Clean Development Mechanism which allows Annex I countries to implement mitigation projects in non-Annex I countries (UNFCCC 2010).

National Conference

The National Dissemination and Consensus-Building Conference, held in Baku in October 2012, provided another opportunity to consult with Azerbaijan's experts to identify the highest priority adaptation and mitigation options at both the national and agricultural region level. The overall program included a detailed presentation of the technical and farmer consultation findings (as outlined in this report), and a half-day consensus-building exercise among participants, with region-focused small group discussions to provide rankings and information for the multi-criteria assessment calculations.

The small groups were given tables that summarized the results of the completed B-C analysis, expert assessment, win-win assessment, and mitigation assessment. The process had a three-part agenda: (1) rank the actions/policies for the focus region from the provided table in order of importance, including crossing off any options that are not relevant, identifying other actions or policies that should be considered, and ranking the resulting overall set of options; (2) rate the importance of three technical criteria by allocating 100 total points across: (a) B-C analysis (net economic benefit), (b) potential to help with or without climate change (win-win potential), and (c) GHG mitigation potential, to reflect the relative importance the group places on achieving each objective; and (3) report back on findings to the full conference in plenary session.

Table 4.8 Ranking of Adaptation Measures for Azerbaijan's Agricultural Regions

Adaptation measure	Specific focus area	Ranking of measure by group				
		National	Irrigated	High rainfall	Low rainfall	Subtropical
Improve farmer access to agronomic technology and information	Fertilizers, herbicides, seed varieties; more efficient use of water	1				
Increase the quality, capacity, and reach of extension services	Demonstration plots	2				
Improve farmer access to hydro-meteorological capacity	Short-term temperature and precipitation forecasts	3				
Create crop insurance program	Promotion of investments in agricultural crops susceptible to drought and hail	4				
Improve irrigation water availability	Rehabilitate irrigation capacity		1	3	3	3
Optimize agronomic practices	Increase and improve fertilizer application			1	1	1
Improve crop varieties	Drought-tolerant varieties		4	2	2	2
Research and improve livestock nutrition, management, and health	Include research on sheltering techniques			5	5	4
Optimize and/or improve irrigation techniques	Sprinkler, drip irrigation		5		4	
Rehabilitate drainage systems and/or improve drainage canals				4		
Create large-scale farms	Farm consolidation		2			
Establish agribusinesses	Assist with corresponding business plans		3			
Improve or introduce pasture management and animal husbandry					6	

Note: Items without entries were not ranked by those groups.

Rankings of the groups, as reported from the conference, are presented in table 4.8. The national group focused on national-scale policies, and as a result presented an entirely different focus from the region-focused groups. The region-focused groups provided additional measures for consideration unique to their regions, and included in their priority lists different numbers of measures (four to six total). Across the regions, there was broad support for improving irrigation water availability, optimizing agronomic practices, and improving crop varieties.

Final Menu of Recommended Adaptation Options

The final menu of recommended adaptation options for Azerbaijan reflects multiple lines of quantitative and qualitative analysis of potential net benefits, including evaluations and recommendations from farmers, stakeholders, and other experts. These measures were identified as important both at the national conference and at the farmer workshops. As indicated in table 4.8, these measures address the climate change risks and corresponding impacts on agriculture. The four national-level measures focused on the following areas:

- **Increase the capacity and reach of extension services.** The capacity and effectiveness of existing extension services may be improved by: (1) providing extension agents with up-to-date information and the necessary means to provide services at the required scale, coverage, and quality and (2) using a wide range of extension methods, including farmer meetings, training courses, exposure visits, farmer-to-farmer extension, demonstrations, and mass media. The economic analysis suggests that the benefits of improving extension services are very likely to outweigh the estimated costs. However, it should be noted that lack of access to resources and the inefficient operation of complementary agricultural services will seriously constrain the impact of extension.

- **Ensure that farmers have access to good quality hydrometeorological information.** The need for better local capabilities for hydrometeorological data, particularly for short-term temperature and precipitation forecasts, is substantial in Azerbaijan. These capabilities are acutely needed to support better farm-level decision-making such as irrigation scheduling, developing an early warning for upcoming extreme events, such as frost, and effective pest and disease forecasting for optimum chemical use. Improved applications of weather and climate information using an integrated and coordinated approach will help to increase and sustain agricultural productivity and to reduce production cost at the farm-level. The B-C analysis of a relatively modest hydrometeorological investment, which includes training and annual operating costs, suggests that benefits of such a program are very likely to outweigh costs.

- **Investigate options for crop insurance, particularly for drought.** Crop insurance programs as one of the tools for risk management also have the potential to contribute toward food security at the individual household level in times of unfavorable weather catastrophes. In stakeholder consultations undertaken for this study, farmers were eager to explore insurance options. However, both due to the cost of subscribing to such insurance programs and the extent of expertise required for their operation, these programs are not expected to be viable for the vast majority of agricultural producers in Azerbaijan. One possible way to expand coverage might be to pilot a privately run weather index-based insurance program. This approach has many potential advantages over traditional multiple-peril crop insurance, including simplification of the product, standardized claim payments to farmers in a district based on the index, avoidance of individual farmer field assessment, lower administrative costs, timelier claim payments after loss, and easier accommodation of small farmers within the program. The drawback of an index-based approach may be the inability to readily insure coverage of damage from pests. In addition, pilot insurance schemes based on weather indices have encountered low demand in many locations, partly because poor farmers are cash and credit constrained; therefore they cannot afford premiums to buy insurance that pays out only after the harvest (Binswanger-Mkhize 2012). Poorly designed insurance schemes may also slow autonomous adaptation by insulating farmers from climate-induced risks. In general, countries may need to first consider improving market access and reducing credit constraints, in order to better create enabling conditions suitable for crop insurance to be effective.

• *Improve farmers' access to rural finance to enable them to access new technologies.* Farmers could acquire technologies through well-targeted and affordable credits to improve crop and livestock yields. However, the current rural finance system with its relatively high interest rate combined with stringent collateral requirements and limited outreach prohibits access to credit for many rural households in Azerbaijan despite the demand. The commercial banks and non-bank financial institutions (NBFIs) need to fine-tune their loan products to the specificities of rural investments: reduce the periodicity of cash-flow, provide longer maturity to match the specific crop and livestock production cycles, and make payments non-monthly. The need for tailoring techniques to shifting climatic conditions without harming ecosystems of the country is pressing and urgent.

Recommended Adaptation Options by Agricultural Region

Recommendations for each agricultural region to improve the resilience of Azerbaijan's agricultural sector to climate change are presented in figures 4.5 through 4.8. At the agricultural region and farm levels, high-priority adaptation measures include improving and/or augmenting irrigation infrastructure; optimizing application of irrigation water at the farm level; and providing more climate-resilient seed varieties along with focused training on how best to cultivate these new crops effectively.

Irrigation water shortages appear likely to occur under climate change—and even if climate does not change in the future, a shortage can occur from competition with growing demand from nonagricultural water users—but these shortages can be addressed through a range of adaptive measures. For example, improvements in farmer trainings could help ensure more efficient on-farm water use during dry seasons, and additional investment in the current irrigation infrastructure could help make better use of available water resources in the agricultural sector. The study's economic analysis suggests that the benefits of these investments would likely exceed the construction costs under most scenarios.

Limitations of the Study

Finally, due to its broad scope, this study necessarily involves significant limitations. These include the need to make simplifying assumptions about many important aspects of agricultural and livestock production in Azerbaijan, and the

Figure 4.5 Irrigated Agricultural Region Priority Adaptation Measures

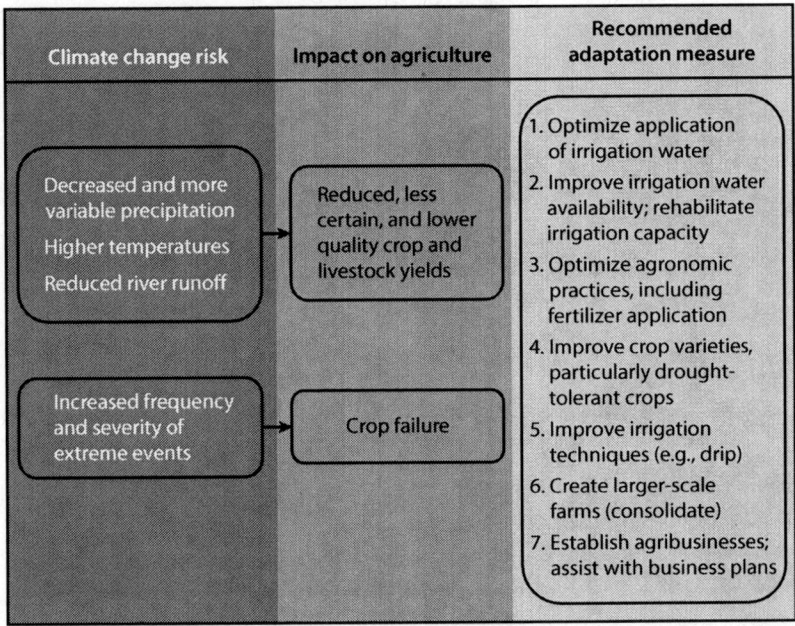

Figure 4.6 High Rainfall Agricultural Region Priority Adaptation Measures

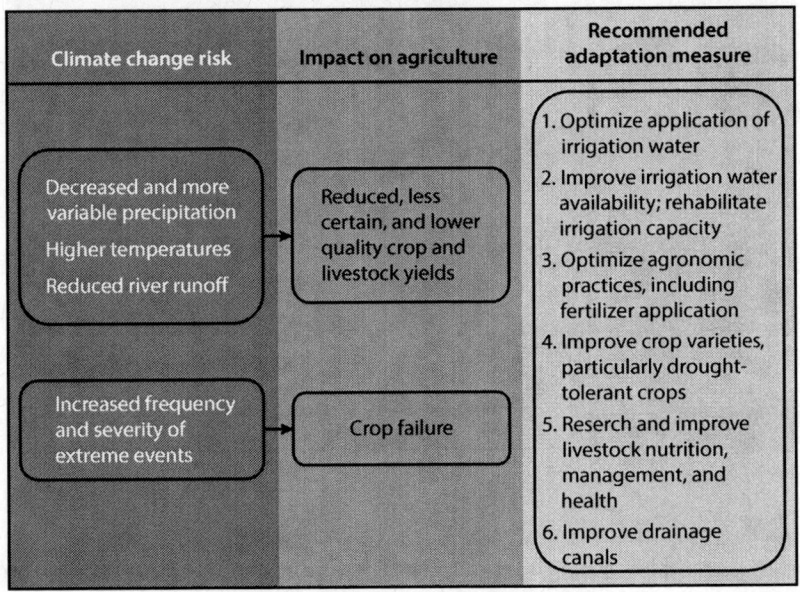

limitations of simulation modeling techniques for forecasting crop yields and water resources. As a result, certain recommendations may require a more detailed examination and analysis than could be accomplished here in order to ensure that specific adaptation measures are implemented in a manner that maximizes their value to Azerbaijani agriculture. However, the authors hope that

Figure 4.7 Low Rainfall Agricultural Region Priority Adaptation Measures

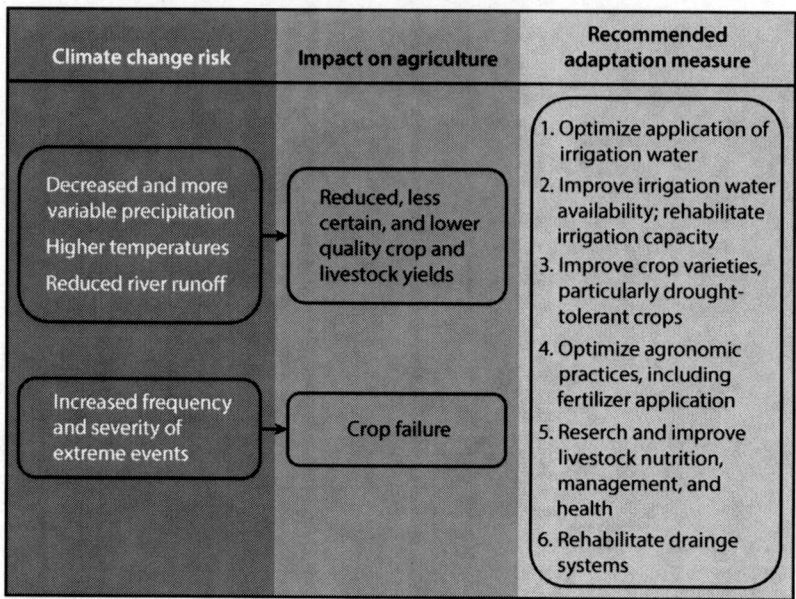

Figure 4.8 Subtropical Agricultural Region Priority Adaptation Measures

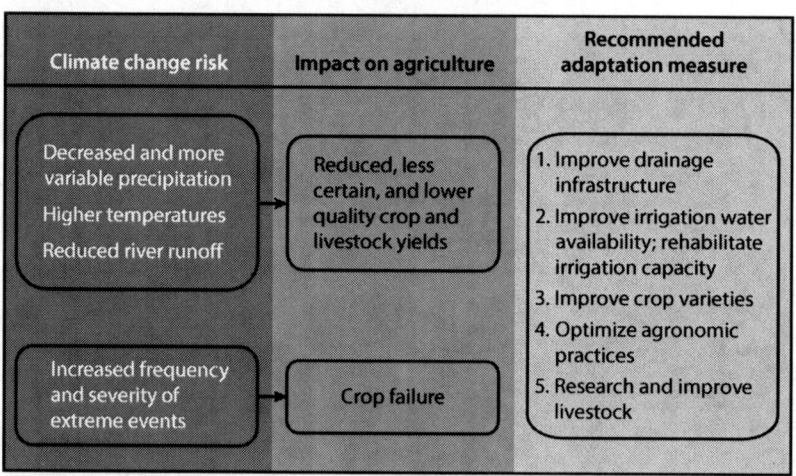

the awareness of climate risks and the analytic capacities built over the course of this study provide not only a greater understanding among Azerbaijani agricultural institutions of the basis of the recommendations presented here, but also an enhanced capability to conduct the required more detailed assessment that will be needed to further pursue the recommended actions.

In addition, it is desirable that the countries of the South Caucasus address climate change through collaboration on issues such as climate-related data sharing and crisis response. There are many challenges to achieving these objectives, but fortunately there is a wide range of existing models of regional-scale institutional arrangements throughout the world, encompassing the scope of regional cooperation for water resources planning, agricultural research and extension, and enhanced hydrometeorological service development and data provision.

References

AZStat (State Statistical Committee of the Republic of Azerbaijan). 2013. State Statistical Committee of the Republic of Azerbaijan website (accessed December 9, 2013), http://azstat.org/indexen.php.

Binswanger-Mkhize, H. P. 2012. "Is There Too Much Hype about Index-Based Agricultural Insurance?" *Journal of Development Studies* 48 (2): 187–200.

Stokes, C. R., S. D. Gurney, M. Shahgedanova, and V. Popovnin. 2006. "Late-20th-Century Changes in Glacier Extent in the Caucasus Mountains, Russia/Georgia." *Journal of Glaciology* 52 (176): 99–109.

UNFCCC (United Nations Framework Convention on Climate Change). 2010. *Second National Communication on Climate Change: A Report under the United Nations Framework Convention on Climate Change.* Republic of Armenia, Ministry of Nature Protection, Yerevan.

Wexler, A. 2013. "Beware the Cotton Glut." *Barrons*, November 16, 2013 (accessed December 9, 2013), http://online.barrons.com/article/SB50001424053111190374750 4579183873733553300.html?mod=BOL_twm_mw.

WWF (World Wildlife Fund Norway, and WWF Caucasus Programme). 2009. "Climate Change in Southern Caucasus: Impacts on Nature, People and Society." Report, WWF Norway, Oslo (accessed October 7, 2013), http://assets.wwf.no/downloads /climate_changes_caucasus___wwf_2008___final_april_2009.pdf.

CHAPTER 5

Georgia: Risks, Impacts, and Adaptation Menu

This chapter summarizes the results of efforts to develop a menu of adaptation options for the agricultural sector in Georgia. It is organized into four sections: (1) climate risk, (2) climate impacts, (3) adaptation assessment, and (4) evaluation and prioritization of adaptation options.

Climate Risk

Historical Climate Trends

The South Caucasus region has seen a variety of changes in climate—increasing temperatures, shrinking glaciers, sea level rise, reduction and redistribution of river flows, decreasing snowfall, and an upward shift of the snowline. In the past 10 years, the region has also experienced more extreme weather events such as flooding, landslides, forest fires, and coastal erosion, resulting in economic losses and human casualties (WWF 2009).

Figures 5.1 and 5.2 present historical temperature and precipitation data for Georgia. Figure 5.1 shows annual temperature and growing season temperatures, 1900–2012. During 1980–2012, average annual temperature increased 1.5°C, while average growing season temperature increased 1.7°C.

In addition to changes in temperature and precipitation, the glaciers are melting rapidly in the region, as they are globally. The volume of glaciers in the Caucasus has been reduced by 50 percent over the last century, and 94 percent of the glaciers have retreated 38 meters per year (Stokes et al. 2006). In Georgia, glaciers are retreating 5–10 meters per year, with a maximum of 25 meters per year. Changes in glacier composition can potentially reduce long-term river flow in Georgia.

Forecasted Changes in Temperature and Precipitation

Georgia has a history of floods and erosion, especially in the last two decades. From 1995 through 2009, floods and erosion, particularly through landslides and mudflow, led to US$650 million in economic losses. Heavy downpours led to

Figure 5.1 Average Annual and Growing Season Temperatures in Georgia, 1900–2012

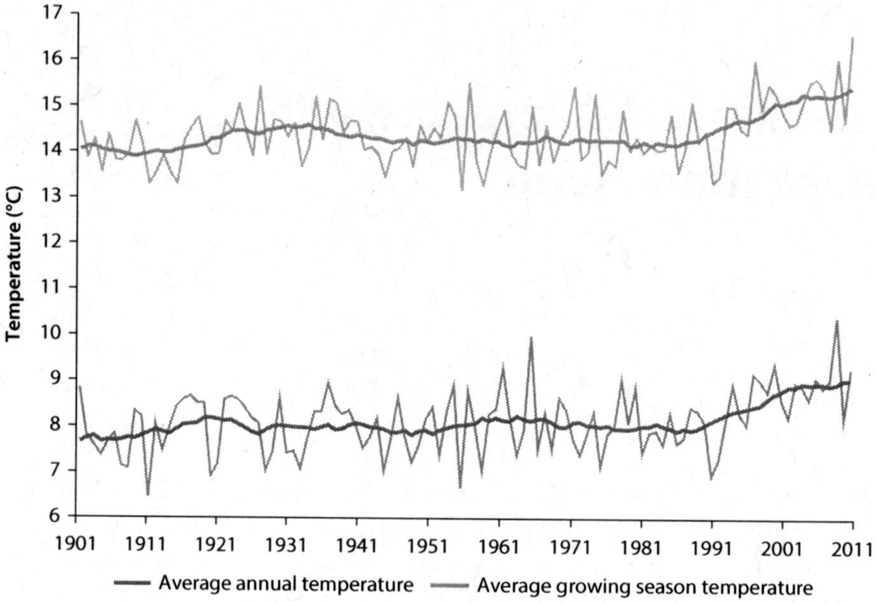

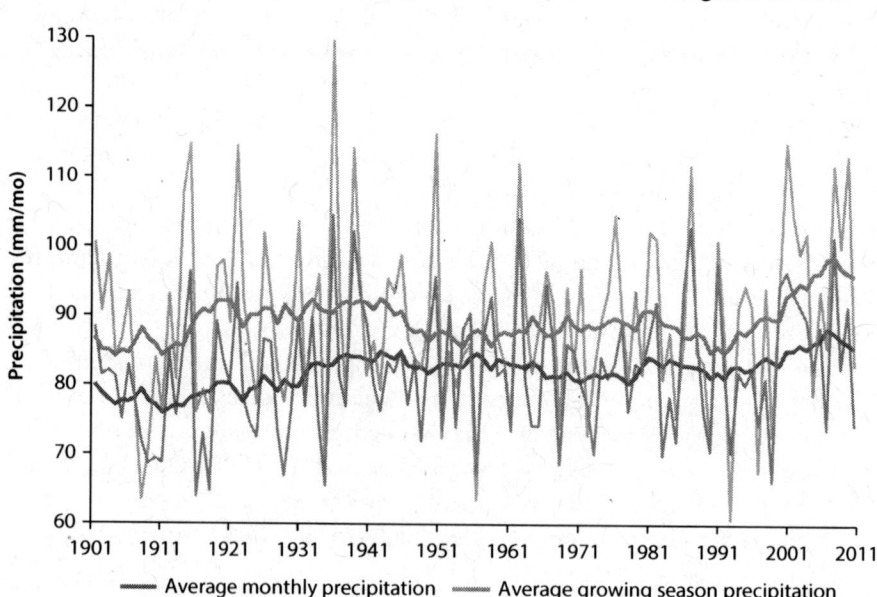

Average annual temperature   Average growing season temperature

Source: University of East Anglia Climatic Research Unit, Norwich, UK.

Figure 5.2 Average Monthly and Growing Season Precipitation in Georgia, 1900–2012

Average monthly precipitation   Average growing season precipitation

Source: University of East Anglia Climatic Research Unit, Norwich, UK.
Note: mm/mo = millimeters per month.

landslides and mudflows in many mountain areas in April–May 2005, resulting in millions of dollars in loss of infrastructure and homes and creating health, sanitation, food, and water problems.

Other large changes in climatic variables are noted in extreme temperature trends. An increasing trend in the number of days per year with maximum temperatures over 25°C was noted in over half of the weather stations monitored. In addition, an increasing number of days with daily minimum temperatures over 20°C were observed in over a quarter of the stations analyzed (UNDP 2011). Floods are reported as killing more people, but drought affects far more people and causes greater economic damages; for example, a severe drought in 2000 affected 700,000 people and caused damages of 5.6 percent of gross domestic product due to its effects on agriculture and on hydropower generation (World Bank 2006). Box 5.1 provides a summary of the recent trends in natural hazards in Georgia.

UNFCCC (2009) includes a case study of the Kvemo Svaneti region, a mountainous area along the central portion of Georgia's northern border in the central agricultural zone, as a region that is most vulnerable to climate change. Disastrous weather events, including floods, landslides, and mud torrents, are becoming more common in this area. Increased frequency and intensity of these phenomena cause land erosion that impacts agriculture, forestry, road transport, and communications. Over the past 50 years, mean air temperature has

---

Box 5.1 Trends in the Natural Hazards in Georgia

Natural hazards pose a serious threat to Georgia, and are estimated to cost the country between US$146 million and $3.3 billion annually. Among the many natural hazards (including earthquakes, mudslides, and avalanches), floods, droughts, and hail storms are of particular concern for the agricultural sector.

Floods are a relatively frequent occurrence in Georgia due to sustained precipitation and rapid snow melt. Flash floods occur in the Kolkheti lowlands, the Caucasus Mountains, and the Meskheti Range. The most disastrous floods took place in 1895, 1922, 1968, and 1987. Droughts have also caused significant damage to Georgia's agricultural sector, and the frequency of severe droughts has increased almost threefold in recent years. Droughts are particularly severe in Shida and Kvemo Kartli, Kakheti, and the upper Imereti regions. In addition, hail storms occur on a seasonal basis throughout the entire territory of the country, though they are most intense and frequent in eastern Georgia. Hail storms destroy approximately 0.7–8 percent of Georgia's agricultural land each year (CENN/ITC 2012).

To address the risks posed by these natural hazards, Georgia established an Emergency Management Department under the Ministry of Internal Affairs in 2005 and the Centre of Monitoring and Prognosis in 2006. These institutions lead disaster risk management activities in the country, which include the provision of food, water, medical service, and temporary shelter and electricity under emergency situations. Figure B5.1.1 shows the duration of droughts (in months) and the number of flood and hailstorm events by year, 1995–2010.

---

box continues next page

Box 5.1 Trends in the Natural Hazards in Georgia *(continued)*

Figure B5.1.1 Disastrous Climatic Events in Georgia, 1995–2010

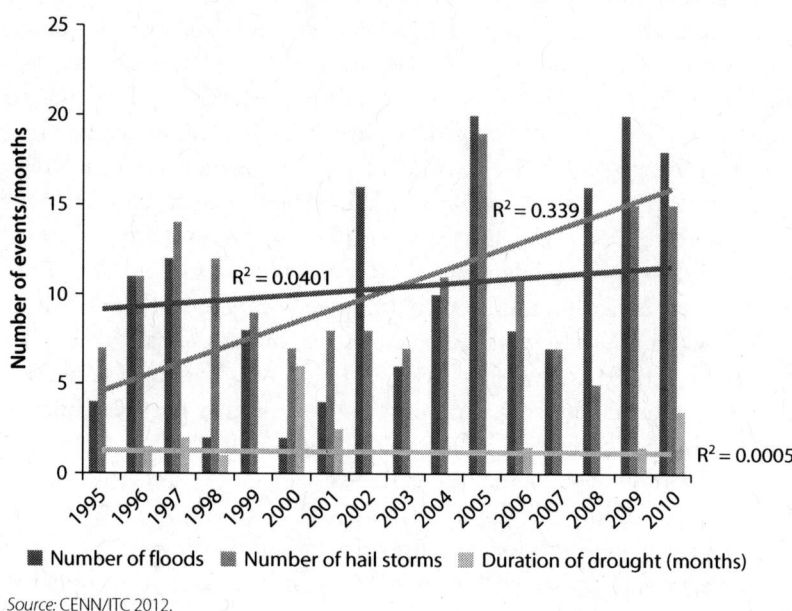

■ Number of floods   ■ Number of hail storms   ░ Duration of drought (months)

Source: CENN/ITC 2012.

risen 0.4°C, and precipitation has increased 106 millimeters (mm), or 8 percent. Increased extreme events are apparent in the frequency of floods doubling from the first half of the period 1967–89, to the second half of that period and, over the same period, floods lasted 25 percent longer. Landslides have increased by 43 percent since 1980, and both mudstreams and droughts have become much more frequent as well.

In addition, the duration and recurrence of droughts have increased in Kvemo Svaneti in the last several decades. The occurrences of pests and diseases in the region's forests, which cover 60 percent of land area, have increased over the past 15–20 years. The Central Caucasus glaciers of this region have decreased in area by 25 percent and by volume from 1.2 to 0.8 cubic kilometers (km³) (UNFCCC 2009). Continued increasing temperatures could cause the glaciers of this region to disappear by 2050.

Forecasted Changes in Temperature and Precipitation

Temperatures in Georgia are predicted to increase for all four agricultural zones, and an increasing frequency of extreme heat events has already been observed in recent years (map 5.1). Although the extent of future warming in Georgia remains uncertain, the overall warming trend is clear. All scenarios associated with this study predict accelerated temperature increases in the country in all four of its agricultural regions—eastern lowlands, western lowlands, eastern mountainous, and western mountainous. The average increase in temperature

Map 5.1 Georgia: Predicted Effect of Climate Change on Average Annual Temperature in the 2040s

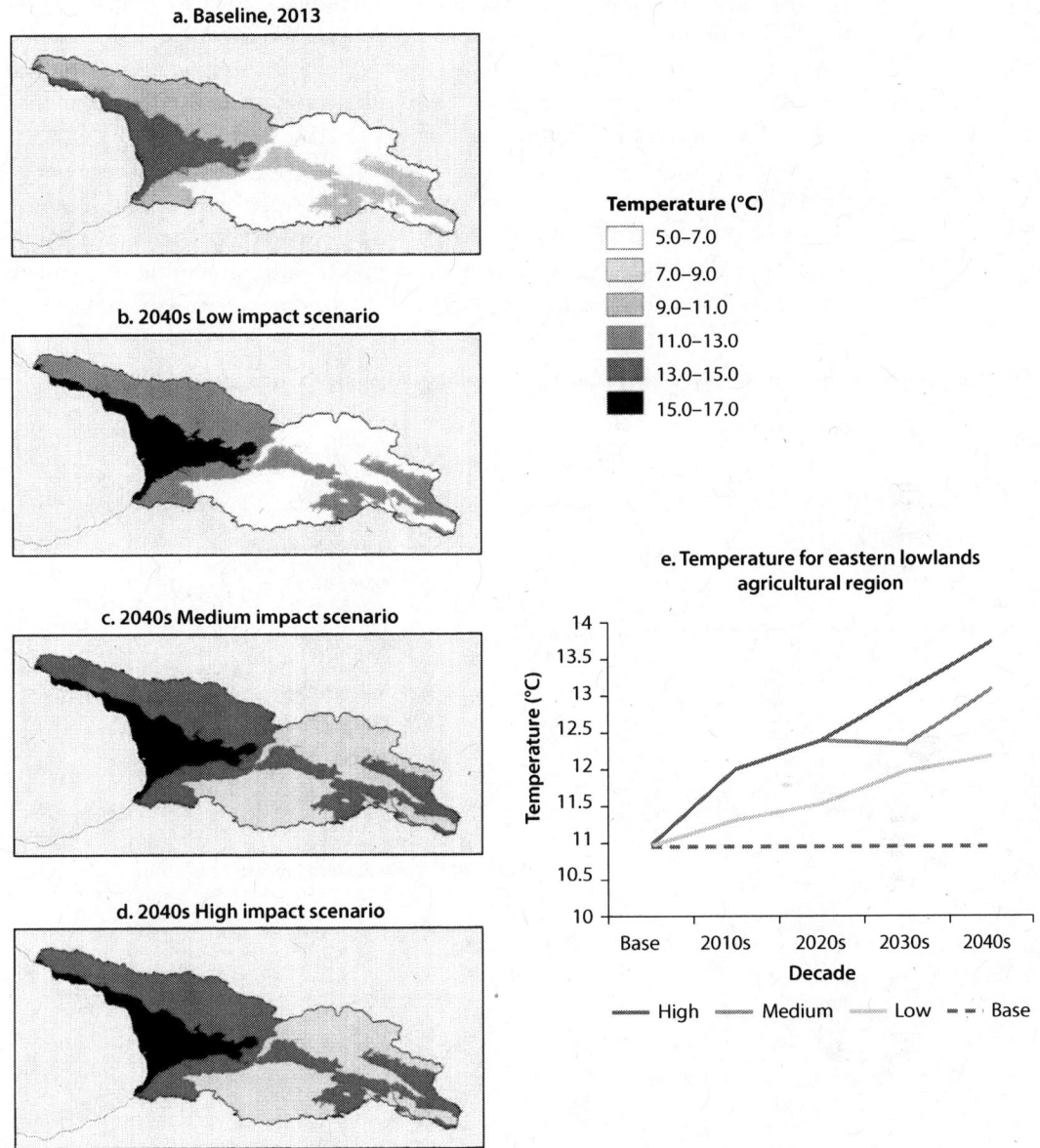

Sources: ©Industrial Economics. Used with permission; reuse allowed via Creative Commons Attribution 3.0 Unported license (CC BY 3.0). Country boundaries are from ESRI and used via CC BY 3.0.

under the Medium Impact Scenario is estimated to be about 2.3°C over the next 50 years, compared with the 0.2–0.4°C increase in temperature observed in the western portion of Georgia and the 0.6°C increase observed in the eastern portion of Georgia over the last 50 years (UNFCCC 2009). Warming could be more modest, but average temperature changes for the Low Impact Scenario nonetheless represent an increase of about 1.2°C, compared to current conditions.

Building Resilience to Climate Change in South Caucasus Agriculture
http://dx.doi.org/10.1596/978-1-4648-0214-0

In all scenarios the warming trend relative to current conditions is about the same magnitude across the four agricultural regions. However, the range of current temperatures across the agricultural regions is quite large. For example, average temperatures in the western lowlands agricultural region are as much as 9°C higher than those in the eastern mountainous region and 3°C higher than those in the western mountainous and eastern lowlands regions.

Estimates of future changes in precipitation in Georgia are much more uncertain than those for temperature. It is unclear whether overall precipitation in Georgia will increase or decrease by 2050, and the extent to which either might occur (map 5.2). Under the Medium Impact Scenario, precipitation remains essentially the same, increasing just 1 mm per year on average. Under the Low

Map 5.2 Georgia: Predicted Effect of Climate Change on Average Annual Precipitation in the 2040s

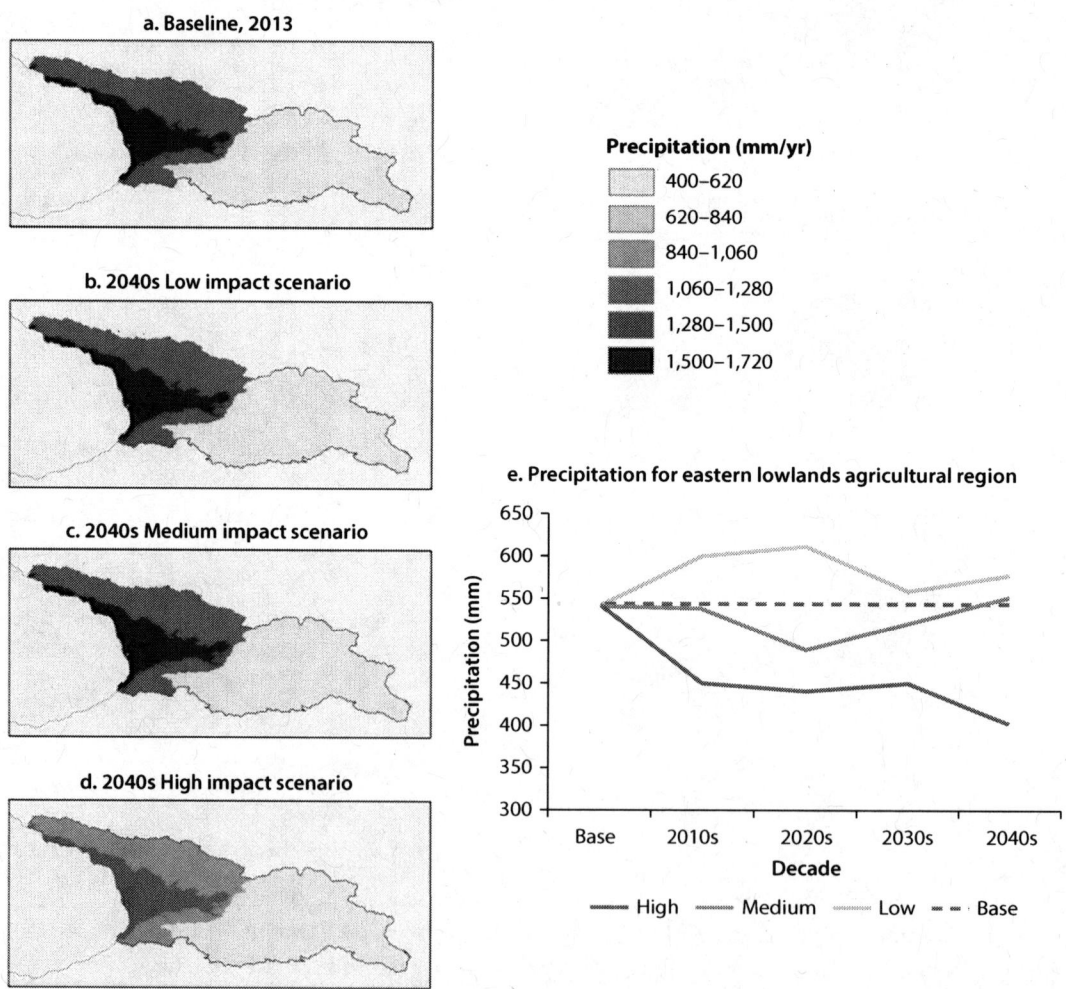

a. Baseline, 2013

b. 2040s Low impact scenario

c. 2040s Medium impact scenario

d. 2040s High impact scenario

Precipitation (mm/yr)
- 400–620
- 620–840
- 840–1,060
- 1,060–1,280
- 1,280–1,500
- 1,500–1,720

e. Precipitation for eastern lowlands agricultural region

— High   — Medium   — Low   — — Base

Impact Scenario, precipitation increases slightly, and under the High Impact Scenario precipitation decreases by 24 percent. Uncertainty at the regional level is even greater, with annual precipitation declines in the western lowlands agricultural region as much as 323 mm.

Predicted temperature increases for Georgia are highest in September and precipitation decreases are greatest in July and August (relative to current conditions). The September temperature increase is estimated to be as much as 5°C in the eastern lowlands agricultural region, when temperatures are already near their highest. Furthermore, forecasted precipitation declines are greatest in the key May–October period, making the late summer and early fall the driest times of year under the High Impact Scenario. Figure 5.3 presents the monthly baseline and forecasted temperatures and precipitation for the eastern lowlands agricultural region.

Between 1995 and 2009, flooding and erosion led to US$650 million in economic losses in Georgia, and climate change is likely to increase the frequency and magnitude of flooding in this region, leading to further damages. While precipitation in Georgia is expected to increase only under the Low and sometimes Medium Impact Scenarios by the 2040s, rainfall events are expected to be more variable, with a high probability of daily to multiday events becoming larger and less frequent. Floods are particularly problematic for Georgia's agriculture sector in the northern plains region near the base of the Greater Caucasus Mountains. Flooding can delay or prevent planting of summer crops in the spring period, and during late summer, flooding can destroy the entire year's growth and prevent timely harvesting. Furthermore, floods in Georgia cause significant loss of agricultural land and rural infrastructure due to destabilized riverbanks. Smaller floods can reduce crop productivity as well through waterlogging stress.

Other issues include drought, frost, high winds, and hail, the last of which has been particularly damaging in recent years (see box 5.1). Emerging literature has indicated that climate change may lead to more frequent and severe hail storms, which could increase hail damage to Georgian agriculture in coming years (Trapp et al. 2007).

Climate Impacts

In order to assess the impact of climate change on the agricultural sector in Georgia, the monthly projections for temperature and precipitation were converted to daily projections for use in crop modeling, as described in chapter 2, box 2.2. The crop models examined the potential effect of climate change on crop yields for seven of Georgia's most important crops, based on a "no adaptation" scenario and not taking into account irrigation water constraints (see table 5.1).

Decline in Crop Yields

As shown in table 5.1, yields of most crops (corn, grape, mandarin orange, potato, tomato, and wheat), both irrigated and rainfed, are expected to decrease in the

Figure 5.3 Georgia: Effect of Climate Change on Monthly Temperature and Precipitation Patterns for the Eastern Lowlands Agricultural Region, 2040s

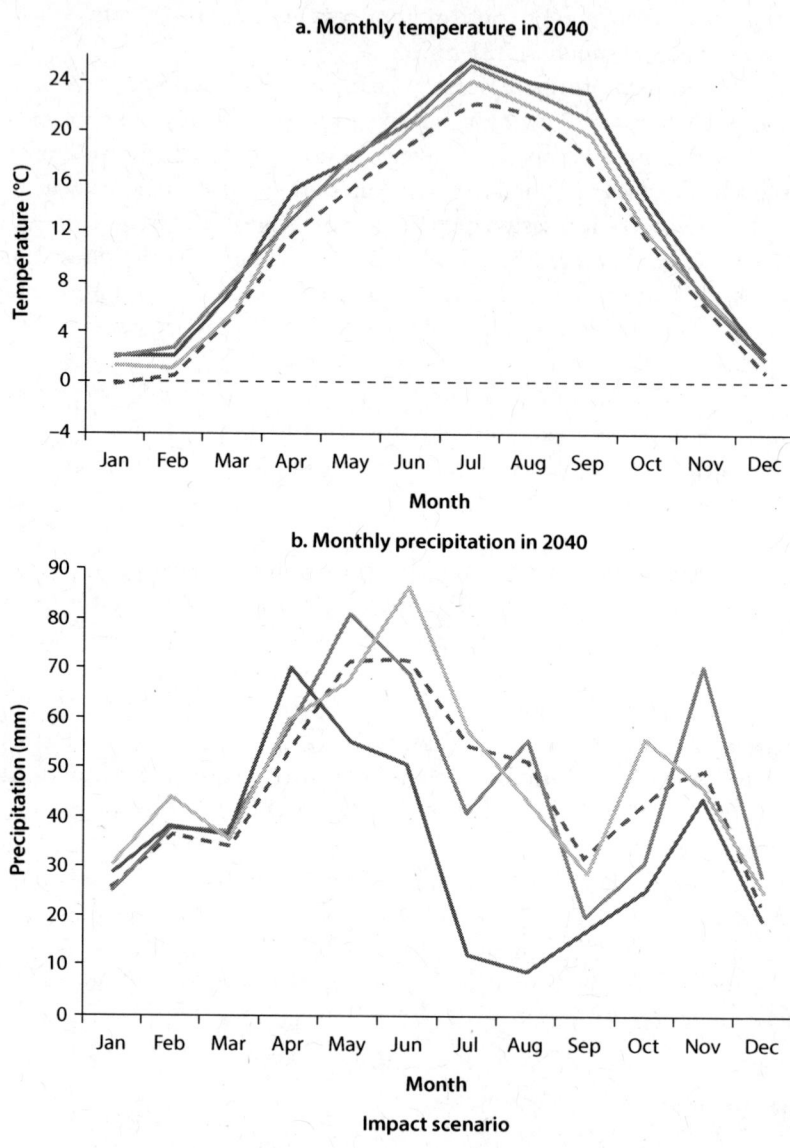

a. Monthly temperature in 2040

b. Monthly precipitation in 2040

Impact scenario

——— High ——— Medium ——— Low – – Base

Source: World Bank data.
Note: mm = millimeters.

eastern lowlands, western lowlands, and western mountainous agricultural regions, while yields of pasture across all four agricultural regions and yields of corn, pasture, tomato, and wheat in the eastern mountainous agricultural region are expected to increase. Pasture yields are expected to nearly double under the Medium and High Impact Scenarios in the eastern mountainous agricultural region. Rainfed tomato would experience the greatest estimated decrease in

Table 5.1 Effect of Climate Change on Crop Yields in the 2040s under the Medium Impact Scenario, no adaptation and no irrigation water constraints

Irrigated/ rainfed	Crop	Change in yield (%)			
		Eastern lowlands	Eastern mountainous	Western lowlands	Western mountainous
Irrigated	Corn	−4	48	−4	−3
	Grape	−5	−5	−5	−5
	Mandarin orange	−5	n.a.	−5	n.a.
	Potato	−5	−5	−5	−5
	Tomato	−6	76	−5	−5
	Wheat	−5	69	−5	−5
Rainfed	Corn	−4	48	−4	−3
	Grape	−6	−5	−5	−5
	Mandarin orange	−5	n.a.	−5	n.a.
	Pasture	26	87	20	44
	Potato	−10	−14	−6	−7
	Tomato	−11	55	−9	−11
	Wheat	−5	69	−5	−5

Source: World Bank data.

Note: Results are average changes in crop yield, assuming no effect of carbon dioxide fertilization. Declines in yield are shown in shades of orange, with darkest representing biggest declines; increases are shaded green, with darkest representing the biggest increases.

n.a. = not applicable (indicates that the crop was not analyzed in that country).

yields, with about 11 percent reductions under the Medium Impact Scenario for both the eastern lowlands and western mountainous regions. As expected, irrigation increased yields and reduced yield variability in the predictions.

Although table 5.1 reflects the assumption that irrigation water will not be constrained, changes in temperature and precipitation resulting from climate change are expected to have an impact on water resources in Georgia. As a result, a more detailed water resource analysis is also needed to determine the extent of climate change impacts. The study team conducted a water availability analysis for Georgia using the Water Evaluation and Planning System (WEAP) and the Climate and Runoff Hydrologic Model (CLIRUN). Next, water supply estimates were matched with forecasts of water demand for all sectors, including agriculture, to determine water availability. Agricultural water demand was estimated using the AquaCrop model (see chapter 2, box 2.2 for more information).

Water Supply Declines, Demand Increases

Figure 5.4 presents the estimated effect of climate change on mean monthly runoff in Georgia in the 2040s. The runoff indicator is directly relevant to agricultural systems and provides insight into the risk of climate change for agricultural water availability, as well as the implications of climate change for water resource management. As shown in figure 5.4, under the High and Medium Impact Scenarios, runoff is expected to decline in the 2040s during the key May–October growing season. Further modeling indicates that irrigation water shortages already occur under the baseline, and they are predicted to rise significantly under climate

Figure 5.4 Estimated Effect of Climate Change on Mean Monthly Runoff Average in the 2040s for all Georgian Basins

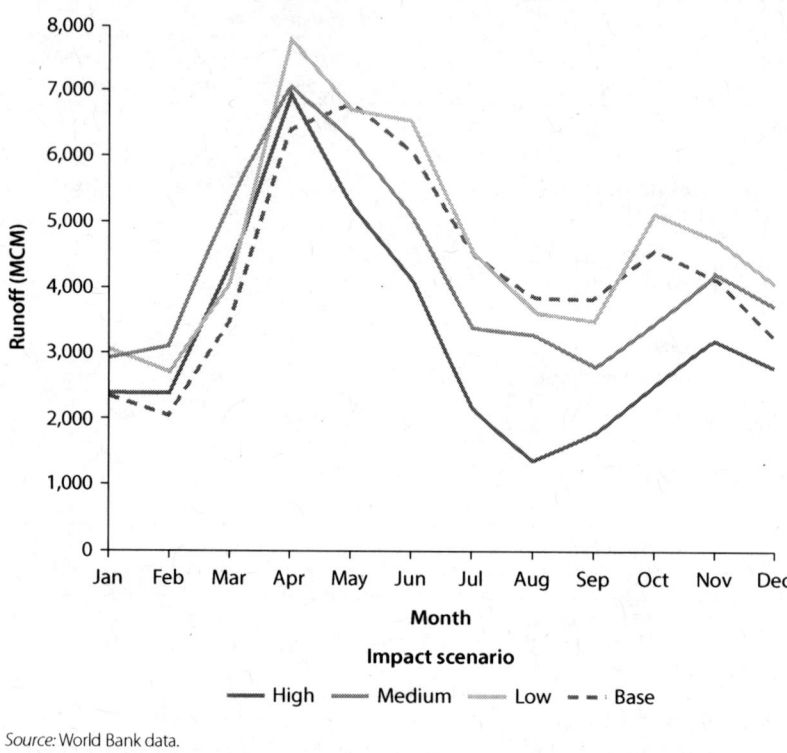

Source: World Bank data.
Note: MCM = million cubic meters.

change, leading to an estimated increase in crop water demand during the summer months when the water supply is most affected by climate change.

Negative Net Climate Effects

Estimates from this study show that Georgian agriculture will be negatively affected by the direct impact of temperature and precipitation changes on crops, the increased irrigation demand required to maintain yields, and the decline in water supply associated with higher evaporation and lower rainfall. Though climate change has a relatively mild negative effect on crop yields when ignoring irrigation water constraints, and is even predicted to boost yields of certain crops in Georgia (particularly in the eastern mountainous region), the phenomenon is expected to greatly decrease irrigation water availability in the country during key months for agricultural productivity. When this decline in water supply is taken into account, projected decreases in crop yields are expected to be much more substantial than those shown in table 5.1. Yield estimates including analysis of predicted changes in water availability are presented in table 5.2.

The direct effects of climate change on livestock production in Georgia could also be severe, but the methods available for quantitatively assessing effects on livestock are relatively untested. There is a robust literature establishing that

Table 5.2 Effect of Climate Change on Irrigated Crop Yields Adjusted for Estimated Irrigation Water Deficits in the Alazani Basin in the 2040s

a. Crop yield impact due to temperature and precipitation changes, without considering irrigation water constraints

	Change in yield (%)	
Crop	Eastern lowlands	Eastern mountainous
Corn	−4	48
Grape	−5	−5
Mandarin orange	−5	n.a.
Potato	−5	−5
Tomato	−6	76
Wheat	−5	69

b. Crop yield impact due to temperature and precipitation changes, as well as forecasted irrigation water constraints

	Change in yield (%)	
Crop	Eastern lowlands	Eastern mountainous
Corn	−33	3
Grape	−30	−30
Mandarin orange	−34	n.a.
Potato	−34	−34
Tomato	−35	23
Wheat	−34	17

Source: World Bank data.
Note: Results are average changes in crop yield, assuming no effect of carbon dioxide fertilization. Declines in yield are shown in shades of orange, with darkest representing biggest declines; increases are shaded green, with darkest representing the biggest increases.
n.a. = not applicable (indicates that the crop was not analyzed in that country).

increases in temperature decrease livestock productivity (Thornton et al. 2009), but suitable modeling tools for quantifying the effect in the Georgian context are not available. According to the analysis in this study, the indirect effect of climate change on livestock feedstocks, including pasture, would be positive, thus providing a counterbalance to the negative direct heat stress effects cited in the literature.

Adaptation Assessment

After examining the local climate risk and likely impacts of climate change on Georgia's agricultural sector, the study team conducted an adaptation assessment of the sector, both at the national and regional levels. This involved stakeholder outreach to elicit information about current farming practices, observing impacts of climate change thus far, and how farmers are currently adapting to these impacts. In addition, the stakeholder outreach sessions allowed the study team to compile an initial list of priority adaptation options based on input from farmers as well as government officials and other local experts. This section describes the findings of the adaptation assessment and the recommended adaptation options from the stakeholder consultations.

Current Regional Adaptive Capacity

To assess Georgia's current regional adaptive capacity, it was essential that the study team inform and consult with a variety of local stakeholders—including farmers and farmers' associations, local government officials, and other local experts—on the predicted impacts of climate change on agriculture and water resources. The team first met with farmers for a one-day stakeholder workshop in Kachreti in April 2012. A second set of farmer consultations was conducted in October 2012 at three locations (Kachreti, Akhaltsike, and Senaki), representing different agricultural regions of Georgia (map 5.3).

At the initial workshop, participants were given an overview of the study and the potential impacts of climate change on crop yields and water availability in Georgia. They were then asked if they had witnessed changes in climate, and if so what they have done, or would do, to mitigate their effects. All confirmed that several of the impacts have been felt on local farms. The stakeholders at the workshop made it clear that, although farmers are becoming more flexible in their response to climate events, their adaptive capacity remains poor due to poorly maintained irrigation and drainage systems, limited financial resources, and inadequate support from and access to the available extension services.

At the subsequent farmer consultations, participants were provided a list of potential climate adaptations. They were asked to remove any irrelevant adaptations and to add any additional options they believed would be effective.

Map 5.3 Locations of Stakeholder Consultations in Georgia

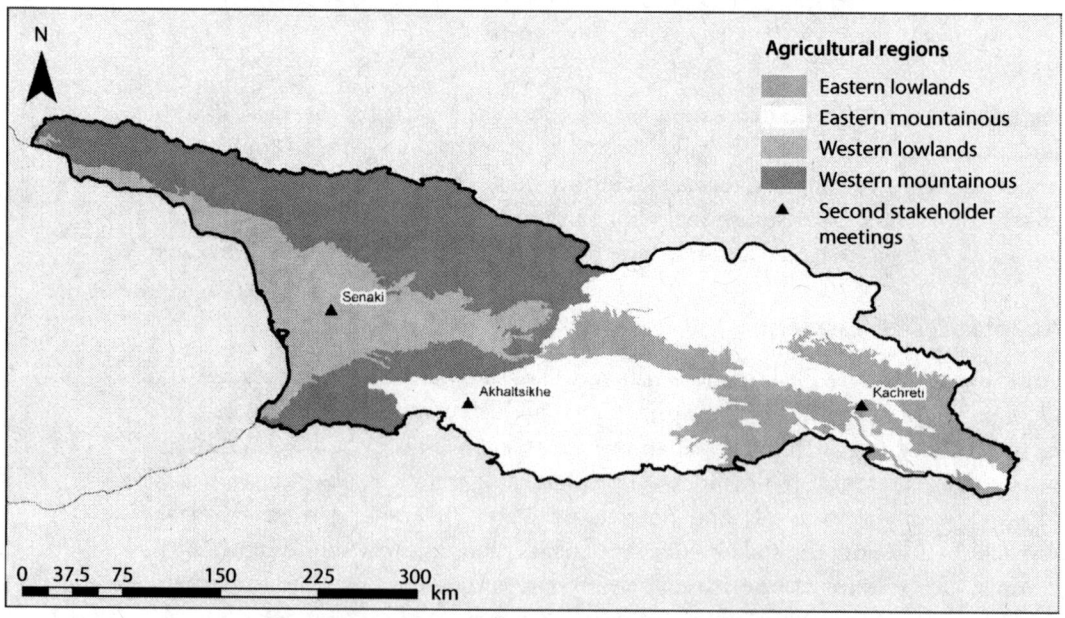

Sources: ©Industrial Economics. Used with permission; reuse allowed via Creative Commons Attribution 3.0 Unported license (CC BY 3.0). Country boundaries are from ESRI and used via CC BY 3.0.
Note: km = kilometers.

Participants then provided rankings for both national-level and regional-level adaptation options. Adaptation rankings varied among regions, reflecting differences in their current climates, topography, and other location-specific factors. The results of this process are reported in the following sections for three of Georgia's four agricultural regions; consultations were not held in the western mountainous agricultural region.

Western Lowlands Agricultural Region: Senaki

This region produces a variety of crops, including citrus, hazelnuts, vegetables, and other orchard crops. The most important weather-related impact noted in this region is hail, which can be problematic during the spring and fall. Flooding can be problematic, but neither frost nor droughts were cited as concerns. Generally, the participants have observed increased weather variability in this region that has made farming more challenging.

Owing to an excess of rainfall, the most important adaptation options to farmers in this region are improving drainage systems and increasing and improving the application of fertilizer, although improving livestock management and wind breaks are also highly ranked alternatives (table 5.3). Rehabilitation of irrigation systems, optimizing the timing of irrigation water application, and construction of small water reservoirs are also key concerns.

Eastern Lowlands Agricultural Region: Kachreti

Key crops grown in the region include field crops (maize, wheat, barley, sunflowers); horticultural crops (grapes, peaches, nectarines, apricots); and vegetables (potato, eggplant, peppers, watermelon). Livestock are also raised in the region, including cattle, sheep, goats, and pigs. The largest issue in the region is drought, although hail and high winds are also of concern. Temperatures have been rising generally, although last year an early winter occurred that caused extensive frost damage.

Table 5.3 Ranked Recommendations from the Senaki Consultation

Adaptation option	Points
Improve drainage systems	31
Optimize agronomic practices (fertilizer)	31
Improve livestock nutrition and shelter	27
Create wind breaks	25
Construct small reservoirs	17
Adjust variety of crops based on elevation	15
Rehabilitate irrigation infrastructure	9
Optimize irrigation water application	7
Use irrigation to prevent frost damage	7
Access to farm equipment	7
Improve crop and livestock varieties	5
Make soil testing available	2

Highly ranked adaptation options (table 5.4) include rehabilitating irrigation systems, increasing and improving the application of fertilizer, optimizing the timing of irrigation water application, and improving wind breaks. Improving livestock nutrition and shelter, improving drainage systems, and several other adaptation options were considered to be important as well.

Eastern Mountainous Agricultural Region: Akhaltsikhe

The major crop in the region is potatoes, but vegetables and grapes are also grown. Participants reported that the primary focus of agricultural activities in the region is livestock. Extreme events of concern include drought and hail events. Frost was reported as also problematic, but it always has been so. Generally, flooding has not been an issue in the region. Participants reported that farming has become more challenging due to increased meteorological variability.

Both rehabilitating irrigation systems and optimizing the timing of irrigation water application received the highest overall scores from the stakeholder groups (table 5.5). Improving livestock nutrition and shelter and optimizing agronomic practices were also highly recommended.

Table 5.4 Ranked Recommendations from the Kachreti Consultation

Adaptation option	Points
Rehabilitate irrigation infrastructure	27
Optimize agronomic practices (fertilizer)	22
Create wind breaks	18
Optimize irrigation water application	18
Improve livestock nutrition and shelter	12
Irrigation to prevent frost damage	10
Provide drainage	8
Adjust variety of crops based on elevation	7
Construct small reservoirs	6
Improve access to farm equipment	5
Make seeds locally available	1

Table 5.5 Ranked Recommendations from the Akhaltsikhe Consultation

Adaptation option	Points
Rehabilitation of irrigation	36
Optimize irrigation water application	28
Optimize agronomic practices (fertilizer)	17
Improve livestock nutrition and shelter	17
Improve crop and livestock varieties	15
Create wind breaks	14
Use irrigation to prevent frost damage	14
Adjust variety of crops based on elevation	12
Improve access to farm equipment	10
Hail nets	6
Local seed production	4
Construct small reservoirs	3

Current National-Level Adaptive Capacity and Responses

Participants in all three regions generally agreed there is a need to improve extension services. This adaptation, along with improving market access and access to long-term, low-interest loans, were the highest-ranked items of the adaptations recommended by farmers (table 5.6). Currently loans are difficult for farmers to obtain and those available are most often short term and high interest.

The second tier of adaptation options reflect the following needs: (1) expand farmer support services to crop insurance, (2) research to identify new crop and livestock varieties, and (3) improve hydrometeorological capacity. While farmers said that crop insurance was sometimes available on the private market, it is often too expensive. They were very interested in securing insurance against losses such as hail and frost.

Climate change clearly challenges Georgia farmers' adaptive capacity. The combination of droughts, frost, hail, and warming is especially disruptive. While the current on-farm adaptation responses have been partially successful, implementation of new programs, policies, and infrastructure investments are needed.

Evaluation and Prioritization of Adaptation Options

The menu of adaptation options to improve the resilience of Georgia's agricultural sector to climate change is derived from the results of stakeholder consultations described in the previous section, in addition to quantitative benefit-cost (B-C) modeling, qualitative analysis, and expert input from international and local teams.

Benefit-Cost Analysis

The study team conducted quantitative B-C analyses for the following nine adaptation options: (1) improving irrigation capacity and efficiency by new investments or rehabilitation to optimize application of irrigation water, (2) improving drainage capacity and efficiency by new investments or rehabilitation, (3) shifting to new crop varieties in irrigated areas, (4) optimizing fertilizer application, (5) improving hydrometeorological services, (6) improving extension services, (7) optimizing basin-level application of irrigation water,

Table 5.6 Stakeholder-Ranked National-Level Climate Adaptations

Adaptation option	Points
Improve extension services	56
Improve market access	51
Provide low-interest, long-term loans to farmers	46
Create crop insurance program	41
Research new crop/livestock varieties	40
Improve hydrometeorological capacity	36
Rehabilitate road infrastructure	2

(8) adding water storage capacity, and (9) installing hail nets for selected crops. The results of the B-C analyses for hail nets and optimizing fertilizer use are presented in tables 5.7 and 5.8 as examples of these analyses.

The tables show the B-C ratios for each crop assessed under the baseline and each climate scenario, using average price assumptions. B-C ratios below 1 indicate that the option is not favorable (that is, costs outweigh benefits). As shown in table 5.7, hail nets are not considered to be a favorable option under any of the scenarios in Georgia's eastern lowlands agricultural region (for grape and tomato) because the costs consistently outweigh the benefits for this option. As shown in table 5.8, optimizing fertilizer use is favorable for several crops (grapes, potato, and tomato) under all scenarios, but is not favorable for corn or wheat.

Table 5.7 Benefit-Cost Ratios for Hail Nets to Protect Selected Crops in Georgia's Eastern Lowlands Agricultural Region

Irrigated/ rainfed	Crop	Climate scenarios			
		Base	Low	Medium	High
Irrigated	Grape	0.03	0.03	0.03	0.03
	Tomato	0.10	0.10	0.20	0.20
Rainfed	Grape	0.03	0.03	0.03	0.03
	Tomato	0.09	0.10	0.10	0.10

Source: World Bank data.
Note: Results are the estimated benefit-cost (B-C) ratios associated with the rehabilitation of irrigation infrastructure, by crop and climate scenario. B-C ratios less than 1 indicate that the costs exceed the benefits.

Table 5.8 Benefit-Cost Ratios for Optimizing Fertilizer Use in Georgia's Eastern Lowlands Agricultural Region

Irrigated/ rainfed	Crop	Climate scenarios			
		Base	Low	Medium	High
Irrigated	Corn	0.4	0.5	0.5	0.5
	Grape	10.0	10.0	10.0	10.0
	Pasture	1.0	1.0	1.0	1.0
	Potato	9.0	9.0	9.0	8.0
	Tomato	8.0	9.0	12.0	12.0
	Wheat	0.2	0.2	0.3	0.3
Rainfed	Corn	0.4	0.5	0.5	0.5
	Grape	10.0	10.0	10.0	10.0
	Pasture	1.0	1.0	1.0	1.0
	Potato	8.0	8.0	8.0	7.0
	Tomato	7.0	8.0	10.0	9.0
	Wheat	0.2	0.2	0.3	0.3

Source: World Bank data.
Note: Results are the estimated benefit-cost (B-C) ratios associated with the optimization of irrigation water application, by crop and climate scenario. B-C ratios greater than 1 (shaded in green, with darkest representing the biggest increases) indicate that the benefits of the adaptation measure exceed the costs, while B-C ratios less than 1 (no shading) indicate that the costs exceed the benefits.

Assessment of GHG Mitigation Potential of Adaptation Options

Many of the adaptive measures recommended above also yield co-benefits in the form of climate change mitigation. Adaptive practices can significantly reduce nitrous oxide and methane emissions. Nitrous oxide emissions are largely driven by fertilizer overuse, which increases soil nitrogen content and generates nitrous oxide. By improving fertilizer application techniques, nitrous oxide emissions can be reduced while maintaining crop yields, specifically through more efficient allocation, timing, and placement of fertilizers. Mitigation of methane emissions, on the other hand, is largely enabled by increasing the efficiency of livestock production. Optimizing breed choices, for example, serves to increase productivity thereby reducing overall methane emissions. Alternative uses of animal manure (for example, biogas production) and improved feed quality quickens digestive processes, resulting in reduced methane emissions. Finally, adaptive measures such as conservation agriculture and manual weeding may also reduce the emissions associated with agricultural production and by heavy machinery use. Similarly, increased irrigation efficiency reduces energy required to pump groundwater.

The potential for adaptive agricultural practices to simultaneously mitigate climate change has already garnered attention in Georgia. As a transition country (UNFCCC Non-Annex 1, that is, not obligated by greenhouse gas [GHG] emissions caps), Georgia has submitted two National Communications to the United Nations Framework Convention on Climate Change (UNFCCC 2009), and some of the Georgia Government's agricultural policies address adaptation and mitigation priorities in the agricultural sector. Some mitigation projects in Georgia are already under way.

The World Bank's Agricultural Research, Extension, and Training Project, now complete, disseminated agricultural knowledge to increase sustainable agricultural production and reduce pollution of natural resources. Specifically, the project mitigated climate change through the adoption of 200 biogas digesters that reduced methane emissions and timber use.

In addition, an afforestation project with hazelnut plantations in western Georgia is under way through Agrigeorgia, LLC, Georgia, and GET-Carbon USA, working with communities in the Samegrelo region of Georgia. The project aims to reclaim abandoned lands in the sustainable production of food that can be sold locally and internationally, to increase employment and technology transfer to local communities, and to use carbon finance to increase economic returns and reduce risk. The project is scheduled to last from 2007 through 2057, involve 250 households, and could have a benefit of 300,000 tons of carbon dioxide mitigation.

National Conference

The National Dissemination and Consensus-Building Conference, held in Tbilisi in October 2012, provided another opportunity to consult with Georgia's experts to identify the highest priority adaptation and mitigation options at both the national and agricultural region level. The overall program included a detailed

Building Resilience to Climate Change in South Caucasus Agriculture
http://dx.doi.org/10.1596/978-1-4648-0214-0

presentation of the technical and farmer consultation findings (as outlined in the last section), and a half-day consensus-building exercise among participants, with region-focused groups providing rankings and information for the multi-criteria assessment calculations.

The small groups were presented with tables that summarized the results of the completed B-C analysis, expert assessment, win-win assessment, and mitigation assessment. The agenda for the process was in three parts: (1) rank the actions/policies for the focus region from the provided table in order of importance, including crossing off any options that are not relevant, identifying other actions or policies that should be considered, and ranking the resulting overall set of options; (2) rate the importance of three technical criteria by allocating 100 total points across: (a) B-C analysis (net economic benefit), (b) potential to help with or without climate change, and (c) GHG mitigation potential, to reflect the relative importance the group places on achieving each objective; and (3) report back on findings to the full conference in plenary session.

Rankings of the groups, as reported in the conference, are presented in table 5.9. The national group focused on national-scale policies, while the regional group provided additional measures for consideration unique to their regions. In the regional group participants showed broad support for improving irrigation water availability, optimizing agronomic practices, and improving crop varieties. One group considered measures for both the eastern and western mountainous regions, because of the close similarities between these areas in Georgia.

Final Menu of Recommended Adaptation Options

The final menu of recommended adaptation options for Georgia reflects multiple lines of quantitative and qualitative analysis of potential net benefits, including evaluations and recommendations from farmers, stakeholders, and other experts. These measures were identified as important both at the national conference and at the farmer workshops. The following seven items are recommended for adoption at the national level (figure 5.5):

- *Improve farmer access to agronomic technology and information.* Through improved extension services, farmers could access technologies to improve crop yields—for example, obtaining new seed varieties or investing in drip irrigation. More targeted and practical trainings, such as demonstration plots, could lead to the use of better technologies and agronomic practices.

- *Improve the quality, capacity, and reach of the extension service, both generally and for adapting to climate change.* There was broad agreement among those surveyed that the capacity of the existing extension and research agencies must be improved to support agronomic practices at the farm level, including implementation of more widespread demonstration plots and increased access to better information on the availability and best management practices of high-yield crop varieties. The study's economic analysis suggests that expansion of extension services is very likely to yield benefits in excess of estimated costs.

Table 5.9 Ranking of Adaptation Measures for Georgia's Agricultural Regions

Adaptation measure	Specific focus area	National	Eastern lowlands	Eastern & western mountainous	Western lowlands
Target research and development to climate risks	Locally relevant agricultural research	1			
Increase the quality, capacity, and reach of extension services	Demonstration plots, training, education	2			
Improve farmer access to hydro-meteorological capacity	Short-term temperature and precipitation forecasts	3			
Improve market access	Link markets, market development	4			
Create crop insurance program	Promote investments in agricultural crops susceptible to drought and hail	5			
Improve intersectoral and interagency coordination in planning and implementation		6			
Improve irrigation water availability	Rehabilitate irrigation capacity		3	3	
Optimize agronomic practices	Increase and improve fertilizer application		1	1	2
Improve crop varieties	Introduce drought-tolerant varieties		2	2	
Research and improve livestock nutrition, management, and health	Include research on sheltering techniques		4	1	4
Optimize and/or improve irrigation techniques	Sprinkler, drip irrigation				3
Rehabilitate drainage systems and/or improve drainage canals					1
Undertake reforestation	Include mixed farming			4	

Note: Items without entries were not ranked by those groups.

- *Improve capacity of hydrometeorological institutions.* Farmers noted the need for better local capabilities for hydrometeorological data, particularly for short-term temperature and precipitation forecasts. Those capabilities are acutely needed in the short term to support better farm-level decision making. The economic analysis of the costs and benefits of a relatively modest hydrometeorological investment, which includes training and annual operating costs, suggests that benefits of such a program are very likely to exceed costs.

- *Improve access to local markets.* In Georgia, a large portion of farmers are involved in subsistence and semi-subsistence farming and are frequently exposed to marketing problems. More must be done to improve markets if the agricultural sector potential would be realized. However, it is also clear that without improvements on the producer side, issues related to marketing can

Building Resilience to Climate Change in South Caucasus Agriculture
http://dx.doi.org/10.1596/978-1-4648-0214-0

Figure 5.5 National-Level Priority Adaptation Measures for Georgia

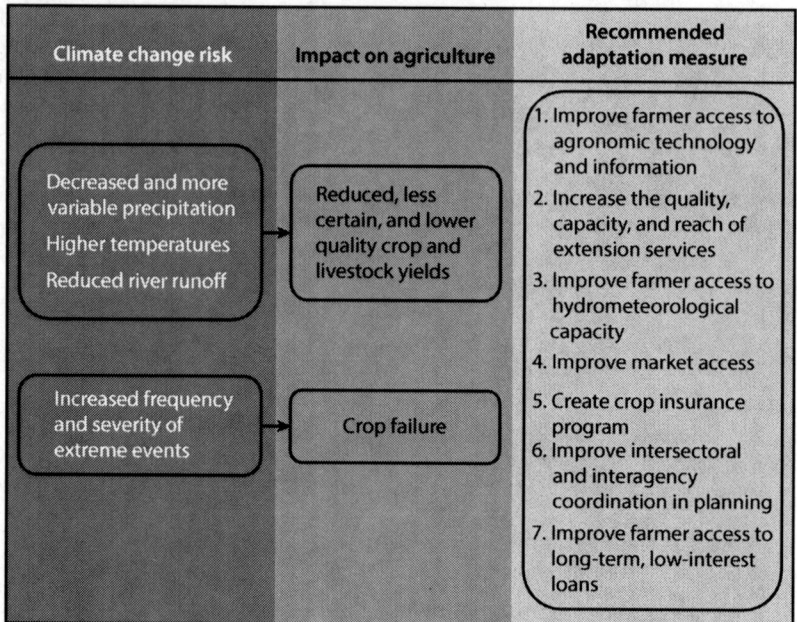

only partially be solved. Efforts should be made to stabilize semi-subsistence farmers' erratic marketing links by providing support to developing their knowledge and skills to produce surplus and in good quality, to support local cooperatives where feasible, and to provide better access to cold storage to facilitate better timing of produce delivery to market.

- ***Investigate options for crop insurance, particularly for drought.*** Crop insurance is not viable for the vast majority of agricultural producers due to its high cost, but farmers remain eager to explore insurance options. One possible way to expand coverage might be piloting a privately run weather index-based insurance program. This approach has many potential advantages over traditional multiple-peril crop insurance, including simplification of the product, standardized claim payments to farmers in a district based on the index, avoidance of individual farmer field assessment, lower administrative costs, timelier claim payments after loss, and easier accommodation of small farmers within the program. The drawback of an index-based approach may be the inability to readily insure coverage of damage from pests. In addition, pilot insurance schemes based on weather indices have encountered low demand in many locations, partly because poor farmers are cash and credit constrained; therefore they cannot afford premiums to buy insurance that pays out only after the harvest (Binswanger-Mkhize 2012). Poorly designed insurance schemes may also slow autonomous adaptation by insulating farmers from climate-induced risks. In general, countries may need to first consider improving market access and reducing credit constraints in order to better create enabling conditions suitable for crop insurance to be effective.

- *Improve intersectoral and interagency coordination and planning.* At the national conference, national institutional stakeholders themselves noted that multiple sectors and agencies are not coordinated in their approach to the agricultural sector. Ideally, government expertise in agronomy, irrigation, hydrometeorology, environmental concerns, subsidy policy, marketing, and rural finance and development can be coordinated to enhance the climate resilience of the agricultural sector to improve the current situation and prepare for future challenges of climate change.

- *Improve farmers' access to rural finance to enable them to access new technologies.* Farmers could acquire technologies through well-targeted and affordable credits to improve crop and livestock yields. However, the current rural finance system, with its relatively high interest rate combined with stringent collateral requirements and limited outreach, prohibits access to credit for many rural households despite the demand. The commercial banks and non-bank financial institutions (NBFI) need to tailor their loan products to the specificities of rural investments: reduce periodicity of cash-flow, provide longer maturity to match the specific crop and livestock production cycles, and pay non-monthly payments. The need for tailoring techniques to shifting climatic conditions without harming ecosystems of the country is pressing and urgent.

As indicated in figure 5.5, these measures address the climate change risks and corresponding impacts on agriculture. In addition, they are responsive to a key policy focus area for Georgia that was established early in the stakeholder process. Specifically, as described in box 5.2, wine grape production has been an important aspect of the agricultural sector in Georgia, but production is hindered by the fact that most grapes in Georgia are rainfed. The study's crop modeling efforts revealed that rainfed grapes are likely to experience only moderate impacts due to climate change, but that irrigated grapes might experience more substantial impacts due to irrigation water shortages. However, many of the priority measures identified in the last section have the potential to help mitigate these impacts, including improving farmer access to agronomic technology and information; increasing the quality, capacity and reach of extension services; improving farmer access to hydrometeorological capacity; and creating a crop insurance program (targeted specifically at hail damage).

Recommended Adaptation Options by Agricultural Region
Recommendations for each agricultural region to improve the resilience of Georgia's agricultural sector to climate change are presented in figures 5.6 to 5.9. At the agricultural region and farm level, high-priority adaptation measures

Box 5.2 Policy Focus Area for Georgia: Wine Grape Production

Grapes—in particular, grapes for producing wine—have long been an important agricultural product in Georgia, by some accounts for several thousand years. Georgia offers excellent soil and climatic conditions for wine production, particularly in the traditional wine-producing regions of Kakheti, Imereti, and Shida Kartli. In 2012 Kakheti accounted for roughly half of Georgia's grape production of 144,000 tons, and Imereti another 25 percent. Grape yields in Georgia appear relatively low compared to other countries: average yields are about 4.1 ton/hectare (ha) in Georgia, about 20 percent less than other Eastern European countries and significantly less than the world average of 9 ton/ha.

The lower yields resulted in part because most grapes in Georgia are rainfed. More important, however, is that a high percentage of Georgia's grapes are wine grapes, which are grown for quality rather than high yield. Most wine grapes are used by small-scale "family" producers, while about 20 percent are used by wine-making enterprises. The export market is currently relatively small, and it was severely affected by a ban on Georgian wine imports by the Russian Federation in 2006. The market is currently recovering and has benefited from cooperative efforts by public and private sector interests (Georgian Wine Association 2011).

Crop modeling shows that Georgian rainfed grapes would experience relatively modest declines in yield (about 5 percent) under the Medium Impact Scenario by the 2040s. This decline is largely associated with increases in temperature.

Because the decline is relatively small and most wine grapes are rainfed, the study concluded that wine grape production should not be significantly affected by climate change. In the relatively few areas where wine grapes are irrigated, however, particularly in the Alazani basin, irrigation water shortages may lead to declines in yield of up to 30 percent. However, many of the measures identified as priorities at the national level have the potential to mitigate these impacts, including improving farmer access to agronomic technology and information; increasing the quality, capacity, and reach of extension services; improving farmer access to hydrometeorological capacity; and creating a crop insurance program (targeted specifically at hail damage).

include optimizing fertilizer application; improving irrigation systems; and providing more climate resilient seed varieties and the training to cultivate them effectively for high yields (all agricultural regions). These measures have high B-C ratios (depending on the region and scenario) and are favored by Georgian farmers.

Limitations of the Study

Because its scope was broad, the study necessarily had significant limitations, including the need to make simplifying assumptions about many important aspects of agricultural and livestock production in Georgia, and the limitations of

Figure 5.6 Priority Adaptation Measures for Georgia's Eastern Lowlands Agricultural Region

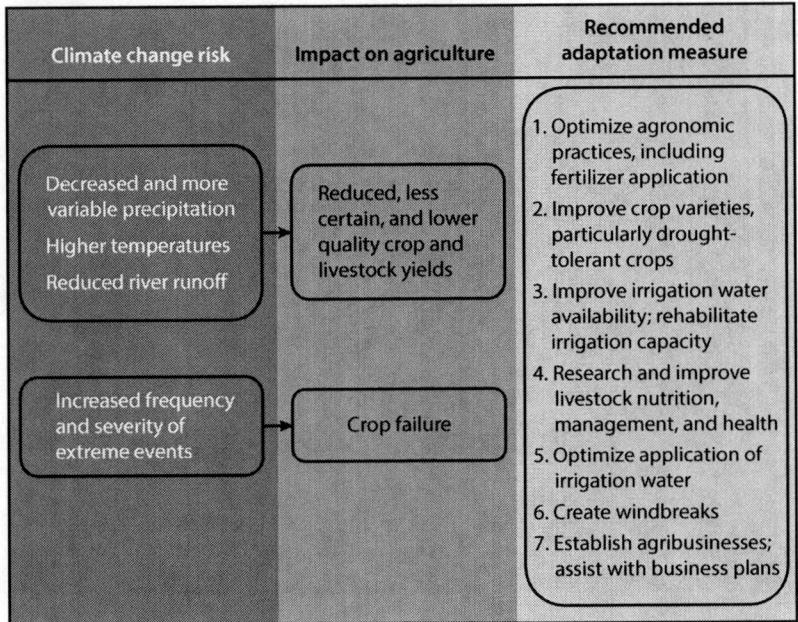

Figure 5.7 Priority Adaptation Measures for Georgia's Western Lowlands Agricultural Region

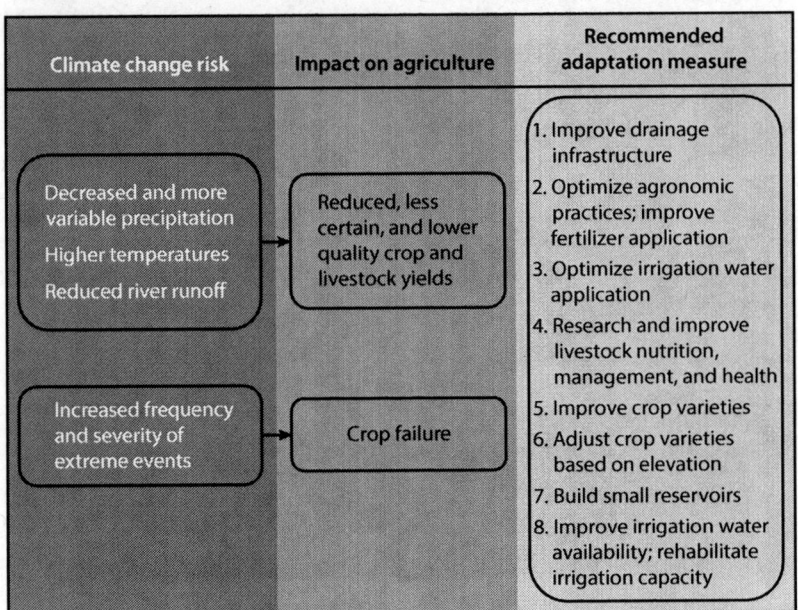

Figure 5.8 Priority Adaptation Measures for Georgia's Eastern Mountainous Agricultural Region

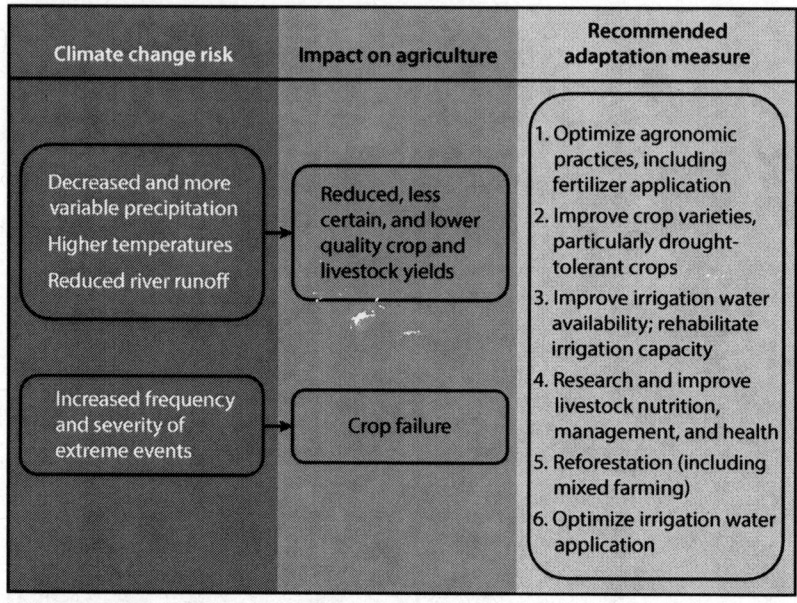

Figure 5.9 Priority Adaptation Measures for Georgia's Western Mountainous Agricultural Region

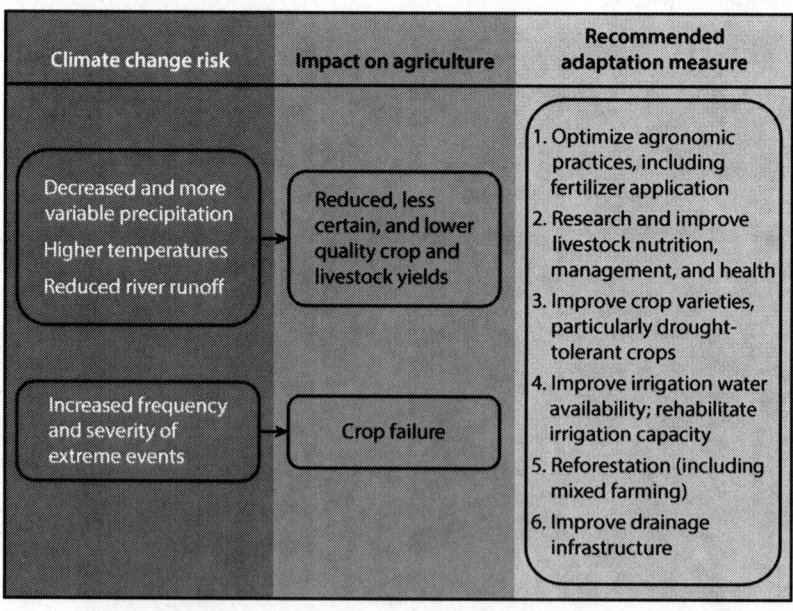

simulation modeling techniques for forecasting crop yields and water resources. As a result, certain recommendations may require a more detailed examination and analysis than could be accomplished in this study in order to ensure that specific adaptation measures are implemented in a manner that maximizes their value to Georgian agriculture. However, the authors hope that the awareness of climate risks and the analytic capacities built over the course of this study provide not only a greater understanding among agricultural institutions on the basis of the recommendations presented here, but also an enhanced capability to conduct the required more detailed assessment that will be needed to further pursue the recommended actions.

In addition, it is recommended that the countries of the South Caucasus address climate change through collaboration on issues such as climate-related data sharing and crisis response. Although achieving these objectives may be subject to many challenges, fortunately there is a wide range of existing models of regional-scale institutional arrangements throughout the world, encompassing the scope of regional cooperation for water resources planning, agricultural research and extension, and enhanced hydrometeorological service development and data provision.

References

Binswanger-Mkhize, H. P. 2012. "Is There Too Much Hype about Index-Based Agricultural Insurance?" *Journal of Development Studies* 48 (2): 187–200.

CENN/ITC (Caucasus Environmental NGO Network/Geo-Information Science and Earth Observation). 2012. "Atlas of Natural Hazards and Risks of Georgia." Paper prepared by CENN and the Faculty of ITC, University of Twente, the Netherlands, as part of the project, "Institutional Building for Natural Disaster Risk Reduction (DRR) in Georgia."

Georgian Wine Association. 2011. "Sector Export Market Development Action Plan (SEMDAP): Wine Sector." Georgian Wine Association, Tbilisi.

Stokes, C. R., S. D. Gurney, M. Shahgedanova, and V. Popovnin. 2006. "Late-20th-Century Changes in Glacier Extent in the Caucasus Mountains, Russia/Georgia." *Journal of Glaciology* 52 (176): 99–109.

Thornton, P. K., J. van de Steeg, A. Notenbaert, and M. Herrero. 2009. "The Impacts of Climate Change on Livestock and Livestock Systems in Developing Countries: A Review of What We Know and What We Need to Know." *Agricultural Systems* 101: 113–27.

Trapp, R. J., N. S. Diffenbaugh, H. E. Brooks, M. E. Baldwin, E. D. Robinson, and J. S. Pal. 2007. "Changes in Severe Thunderstorm Environment Frequency during the 21st Century Caused by Anthropogenically Enhanced Global Radiative Forcing." *Proceedings of the National Academy of Sciences* 104 (50): 19719–23.

UNFCCC (United Nations Framework Convention on Climate Change). 2009. *Georgia:* Second National Communication on Climate Change: A Report under the United Nations Framework Convention on Climate Change. Ministry of Environment Protection and Natural Resources of Georgia and United Nations

Development Programme, Tbilisi, Georgia (accessed November 4, 2013), http://unfccc.int/resource/docs/natc/geonc2.pdf.

UNDP (United Nations Development Programme). 2011. *Regional Climate Change Impacts Study for the South Caucasus Region.* Tblisi: Environment and Security (ENVSEC) Initiative and UNDP (accessed October 17, 2013), http://www.envsec.org/publications/cc_report.pdf.

World Bank. 2006. *Drought Management and Mitigation Assessment for Central Asia and the Caucasus: Regional and Country Profiles and Strategies.* Washington, DC: World Bank (accessed November 4, 2013), http://siteresources.worldbank.org/INTECAREGTOPRURDEV/Resources/CentralAsiaCaucasusDroughtProfiles&Strategies-Eng.pdf.

WWF (World Wildlife Fund Norway, and WWF Caucasus Programme). 2009. "Climate Change in Southern Caucasus: Impacts on Nature, People and Society." Report, WWF Norway, Oslo (accessed October 7, 2013), http://assets.wwf.no/downloads/climate_changes_caucasus___wwf_2008___final_april_2009.pdf.

CHAPTER 6

Climate Change Impacts and Adaptation Options in the South Caucasus Region

The three country studies detailed in chapters 3, 4, and 5 reveal that climate change is already under way in the South Caucasus and that current adaptation measures are insufficient to prevent adverse effects on the agricultural sector. The forecasted impacts of climate change without adaptation measures provide motivation for immediate action at the farm, agricultural region (subnational), national, and multinational (South Caucasus region) levels. Because of the apparent similarities in their agricultural systems, Armenia, Azerbaijan, and Georgia are often considered together for purposes of agricultural planning and management.[1] Therefore, it is prudent to consider climate adaptation measures that could be cooperatively pursued across the national boundaries of these countries that would benefit the region as a whole. To this end, this chapter provides insights on key regional level findings in terms of climate change impacts on the agricultural sector and the evaluation of adaptation options.

Regional Assessment of Climate Risks and Impacts on Crops and Water Resources

Map 6.1 summarizes the effect of climate change on average annual temperature and precipitation across the South Caucasus region by the 2040s under the Medium Impact Scenario. In general, temperatures are expected to increase while precipitation is expected to increase in some areas and decrease in others. In some areas, the combination of these two effects will increase aridity and reduce soil moisture.

- *Temperature changes.* Under the Medium Impact Scenario, temperatures are expected to increase 2.3–2.6°C across the region by the 2040s, considerably greater than the increase of 0.3–0.7°C observed over the past 50 years. Somewhat greater annual temperature increases can be expected in the

Map 6.1 Predicted Effect of Climate Change on Average Temperature and Precipitation in the 2040s under the Medium Impact Scenario

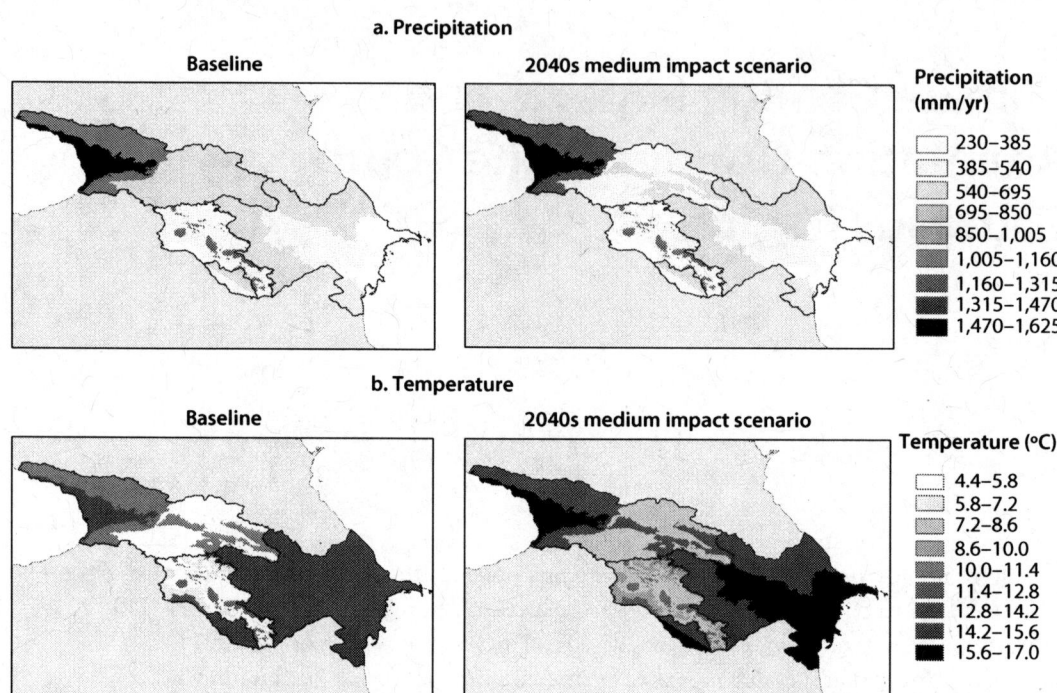

a. Precipitation

Baseline   2040s medium impact scenario

Precipitation (mm/yr)

☐ 230–385
☐ 385–540
☐ 540–695
☐ 695–850
☐ 850–1,005
☐ 1,005–1,160
■ 1,160–1,315
■ 1,315–1,470
■ 1,470–1,625

b. Temperature

Baseline   2040s medium impact scenario

Temperature (°C)

☐ 4.4–5.8
☐ 5.8–7.2
☐ 7.2–8.6
☐ 8.6–10.0
☐ 10.0–11.4
☐ 11.4–12.8
☐ 12.8–14.2
■ 14.2–15.6
■ 15.6–17.0

Sources: © Industrial Economics. Used with permission; reuse allowed via Creative Commons Attribution 3.0. Unported license (CC BY 3.0). Country boundaries are from ESRI and used via CC BY 3.0.
Note: mm = millimeters.

high-elevation areas of western Georgia and Armenia and in the low-elevation areas of Azerbaijan.

- *Precipitation changes.* Precipitation is generally expected to decrease under the Medium Impact Scenario by the 2040s, but the reduction is relatively small (about 50 mm per year on average across the region). The predicted changes in rainfall rates across the three climate scenarios (not shown) vary widely; for example, a modest increase in precipitation is forecasted under the Low Impact Scenario, but there are decreases forecasted under the Medium and High Impact Scenarios. The variation in forecasts at the agricultural region level is quite high; for example, average annual precipitation in Georgia's western lowlands agricultural region is expected to decline by as much as 323 millimeters (mm) per year—six times more than the regional average.

The yearly averages for temperature and precipitation are less important for agricultural production than is the seasonal distribution. Temperature increases are expected to be highest in September, while precipitation decreases are expected to be greatest in July and August (relative to current conditions). The forecasted September temperature increase is as much as 5°C in lower elevation agricultural regions, when temperatures are already near their highest. In addition, forecasted precipitation decreases are greatest in the May–October period critical to agriculture, resulting in the late summer and early fall being the driest times of year under all of the climate scenarios.

Table 6.1 summarizes the impacts of the forecasted changes in temperature and precipitation on crop yields for three low elevation regions of the South Caucasus. The impact on crop yields if no adaptation actions are taken could be severe. For example, under the Medium Impact Scenario, rainfed crop yields may decrease 3–28 percent by the 2040s, while irrigated crops could see more modest yield reductions of 3–16 percent. Of all the crops analyzed, only rainfed pasture is expected to experience increased yields under the Medium Impact Scenario in Georgia's eastern lowlands and Azerbaijan's irrigated agricultural regions.

The crop yield impacts presented in table 6.1 assume availability of sufficient irrigation water; however, a critical factor for irrigated crops is whether the water will be sufficient for adequate irrigation under a changed climate. With increased

Table 6.1 Effect of Climate Change on Crop Yields in the 2040s under the Medium Impact Scenario, no adaptation and no irrigation water constraints

		Change in yield (%)		
Irrigated/rainfed	Crop	Georgia Eastern lowlands	Azerbaijan irrigated	Armenia lowlands
Irrigated	Alfalfa	n.a.	−7	−5
	Apricot	n.a.	n.a.	−5
	Corn	−4	−7	n.a.
	Cotton	n.a.	−3	n.a.
	Grape	−5	−5	−7
	Mandarin orange	−5	n.a.	n.a.
	Potato	−5	−9	−12
	Tomato	−6	n.a.	−16
	Watermelon	n.a.	n.a.	−12
	Wheat	−5	−5	−6
Rainfed	Alfalfa	n.a.	−8	−3
	Apricot	n.a.	n.a.	−28
	Corn	−4	−7	n.a.
	Cotton	n.a.	−13	n.a.
	Grape	−6	−5	−24
	Mandarin orange	−5	n.a.	n.a.
	Pasture	26	5	n.a.
	Potato	−10	−13	−14
	Tomato	−11	n.a.	−19
	Watermelon	n.a.	n.a.	−18
	Wheat	−5	−6	−8

Source: World Bank data.
Note: Results are average changes in crop yield, assuming no effect of carbon dioxide fertilization. Declines in yield are shown in shades of orange, with darkest representing biggest declines; increases are shaded green, with darkest representing the biggest increases.
n.a. = not applicable (indicates that the crop was not analyzed in that country).

temperatures, irrigated crops will require more water to maintain today's yields. In addition, higher temperatures can reduce water runoff, so less surface water is available in rivers for irrigation. Map 6.2 illustrates the mean percentage change in annual water runoff in the 2040s. Map 6.2b shows a widespread reduction in runoff during the critical May–September growing period, when irrigation demands peak.

Forecasts regarding changing water demand (plants requiring more water) and supply (reduced runoff) were used in the Water Evaluation and Planning System water balance model to estimate potential irrigation water shortages under climate change. The results shown in table 6.2 indicate that irrigation water shortages can be expected to occur even without climate change, due to demand from other sectors such as municipal and industrial water users. In addition, the results indicate that irrigation shortages are likely to be amplified under climate change.

The six basins shown in table 6.2 and map 6.3 are those forecasted to have irrigation water shortages in the 2040s under all climate scenarios. These basins

Map 6.2 Mean Percentage Change in 2040s Runoff Relative to Historical Baseline

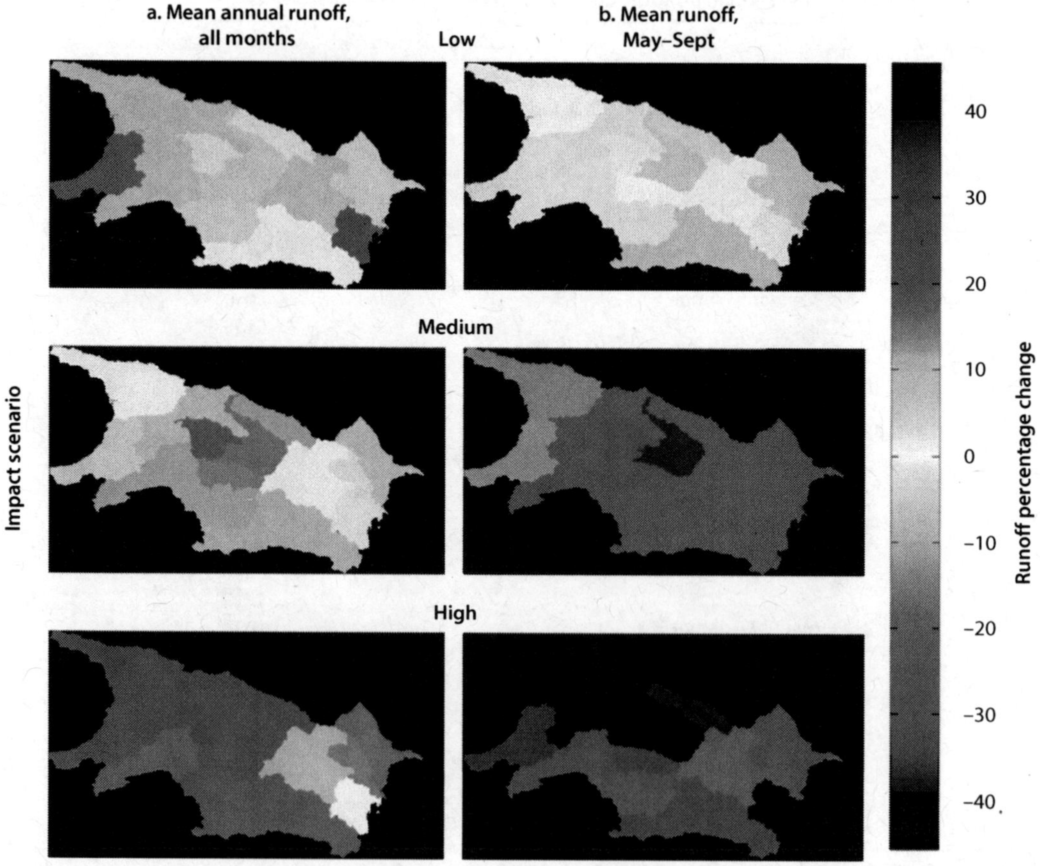

Source: World Bank data.
Note: Scale is percentage change in basin-scale annual runoff quantity from current conditions.

Table 6.2 Effect of Climate Change on Forecasted Annual Irrigation Water Shortfall in the 2040s for Basins with Shortages

River basin	Climate scenario, m³ thousands/ (% irrigation water demand)							
	Base		Low		Medium		High	
Ganikh, Azerbaijan	36.2	(10.5)	43.9	(12.6)	81.5	(23.3)	124.4	(35.3)
Lenkeran/Vilesh/Southern Caspian, Azerbaijan	496.5	(67.6)	523.4	(70.7)	562.2	(75.3)	590.8	(77.5)
Eastern Lower Kur, Azerbaijan	433.0	(67.2)	461.2	(70.2)	498.4	(76.7)	506.9	(78.0)
Samur/Middle Caspian, Azerbaijan	46.6	(5.3)	82.5	(9.3)	197.0	(22.0)	282.6	(30.9)
Upper Araks, Armenia	121.9	(20.6)	140.4	(23.2)	273.3	(44.6)	346.4	(55.4)
Alazani, Georgia	36.5	(14.8)	35.1	(14.1)	75.5	(30.5)	153.7	(61.2)
Total	1170.7		1286.5		1687.9		2004.8	

Source: World Bank data.

Map 6.3 Basins with Forecasted Irrigation Water Shortages by 2050 or Sooner, All Scenarios

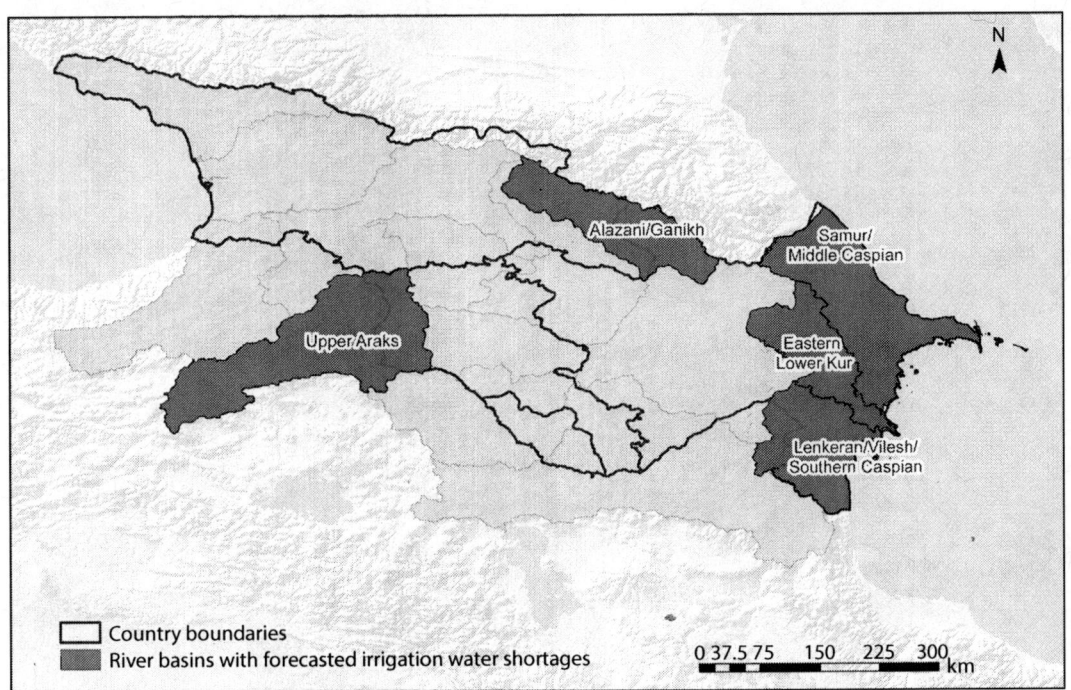

Sources: ©Industrial Economics. Used with permission; reuse allowed via Creative Commons Attribution 3.0 Unported license (CC BY 3.0). Country boundaries are from ESRI and used via CC BY 3.0.
Note: km = kilometers.

include some of those considered most important for agricultural production in the South Caucasus, such as areas where high-value fruit and vegetable crops are grown in Azerbaijan and Armenia, and where wine grapes are grown in Georgia.

The forecasted irrigation water shortages could have a much larger impact on crop yields than might be inferred from table 6.1. If water available for irrigation is insufficient, farmers would be expected to compensate in the short-term by either reducing the cropped area or adjusting to reduced yields as compared to the potential where all irrigation water demands are met and agronomic practices are appropriate. Taking into account the impact of water shortages resulting from climate change, yields of irrigated crops drop further, as shown in table 6.3, resulting in as much as an *80 percent reduction* in yield as a total impact of climate change on the crops—a result that could be

Table 6.3 Effect of Climate Change on Crop Yields in the 2040s Relative to Current Yields for Irrigated Crops, No Adaptation but including Irrigation Water Shortages

	Agricultural region (country/basin), percentage change in yield		
	Eastern lowlands (Georgia)	Irrigate (Azerbaijan)	Lowlands (Armenia)
Crop	Alazani	Southern Caspian	Upper Araks
Baseline (current climate)			
Alfalfa	n.a.	−68	−21
Corn	−15	−68	n.a.
Grape	−13	−57	−18
Potato	−15	−68	−21
Tomato	−15	n.a.	−21
Wheat	−15	−68	−21
Low impact scenario			
Alfalfa	n.a.	−72	−27
Corn	−16	−72	n.a.
Grape	−16	−62	−23
Potato	−18	−72	−28
Tomato	−18	n.a.	−28
Wheat	−17	−72	−27
Medium impact scenario			
Alfalfa	n.a.	−77	−48
Corn	−33	−77	n.a.
Grape	−30	−66	−42
Potato	−34	−77	−51
Tomato	−35	n.a.	−53
Wheat	−34	−77	−48
High impact scenario			
Alfalfa	n.a.	−80	−60
Corn	−64	−79	n.a.
Grape	−55	−68	−53
Potato	−66	−80	−62
Tomato	−67	n.a.	−64
Wheat	−64	−79	−59

Source: World Bank data.
Note: Results are average changes in crop yield, assuming no effect of carbon dioxide fertilization. Declines in yield are shown in shades of orange, with darkest representing biggest declines; increases are shaded green, with darkest representing the biggest increases.
n.a. = not applicable (indicates that the crop was not analyzed in that country).

devastating to the region's agriculture. This fact alone makes the strongest case for immediate action to adapt to a changing climate.

Evaluation of Climate Change Adaptation Measures for the South Caucasus Region

In the South Caucasus region the potential for very large crop yield losses through the 2040s is high and adaptation actions are clearly necessary. Therefore the study team reviewed a large number of measures, including changing crop management practices, improving the capacity of farmers and institutions to better manage resources, and improving irrigation and drainage infrastructure. As detailed in chapters 3, 4, and 5, the study team evaluated these measures using both quantitative analysis and qualitative assessment, relying heavily on input from stakeholders. Each of those three chapters provides details on the results of the benefit-cost (B-C) analyses of certain adaptation options. For these analyses, the costs of a measure (both capital and operating costs) were compared to the benefits of adapting the measure—mainly the economic value of increases in crop yields resulting from the measure relative to a "no adaptation" baseline.

Tables 6.4 and 6.5 provide examples of the B-C analyses conducted to evaluate various adaptation options in each of the three countries. They present the B-C ratios for two adaptation options in Azerbaijan's irrigated agricultural region: rehabilitating irrigation infrastructure and optimizing irrigation water. Table 6.4 shows that for two crops in Azerbaijan's irrigated agricultural region (cotton and potato), rehabilitating irrigation infrastructure is a favorable option (that is, the benefits of this option outweigh the costs under all scenarios). Note that this measure is modeled as a conversion of crops from "rainfed" to "irrigated" status through rehabilitation of currently unusable irrigation infrastructure. However, for four of the crops (alfalfa, corn, pasture, and wheat), the costs of this option

Table 6.4 Benefit-Cost Analysis Results for Rehabilitated Irrigation Infrastructure Where Crops Are Currently Rainfed in Azerbaijan's Irrigated Agricultural Region

Currently rainfed crop	Climate scenarios, estimated benefit-cost ratios			
	Base	Low	Medium	High
Alfalfa	0.40	0.40	0.40	0.40
Corn	0.30	0.30	0.30	0.30
Cotton	4.00	4.00	5.00	6.00
Grape	0.50	0.70	2.00	3.00
Pasture	0.10	0.20	0.20	0.20
Potato	8.00	8.00	9.00	9.00
Wheat	0.03	0.03	0.04	0.03

Source: World Bank data.
Note: Results are the estimated benefit-cost (B-C) ratios associated with the rehabilitation of irrigation infrastructure, by crop and climate scenario. B-C ratios greater than 1 (shaded in green) indicate that the benefits of the adaptation measure exceed the costs, while B-C ratios less than 1 (no shading) indicate that costs exceed benefits.

Table 6.5　Illustrative Benefit-Cost Analysis Results for Optimizing the Application of Irrigation Water in Azerbaijan's Irrigated Agricultural Region

Irrigated/rainfed	Crop	Climate scenarios, estimated benefit-cost ratios			
		Base	Low	Medium	High
Irrigated	Alfalfa	2.00	1.00	2.00	2.00
	Corn	2.00	3.00	3.00	2.00
	Cotton	0.00	0.00	0.00	13.00
	Grape	0.00	0.05	0.20	0.60
	Pasture	0.00	0.02	0.90	1.00
	Potato	6.00	9.00	34.00	73.00
	Wheat	0.50	0.40	0.40	0.50
Rainfed	Alfalfa	2.00	1.00	1.00	2.00
	Corn	2.00	2.00	3.00	2.00
	Cotton	0.00	0.00	0.00	9.00
	Grape	0.00	0.05	0.20	0.50
	Pasture	0.00	0.02	0.50	0.80
	Potato	5.00	7.00	27.00	58.00
	Wheat	0.50	0.40	0.30	0.50

Source: World Bank data.

Note: Results are the estimated benefit-cost (B-C) ratios associated with the optimization of irrigation water application by crop and climate scenario. B-C ratios greater than 1 (shaded in green, with darkest representing the biggest increases) indicate that the benefits of the adaptation measure exceed the costs, while B-C ratios less than 1 (no shading) indicate that costs exceed benefits.

outweigh the benefits, and for one (grape), the option is favorable only under the Medium and High Impact Scenarios.

Table 6.5 presents the B-C ratios for optimizing the application of irrigation water, which is a demand management measure—as opposed to an infrastructural measure, such as rehabilitating irrigation systems. The results show that this option is favorable for four crops—irrigated corn and potato and rainfed corn and potato—because the benefits outweigh the costs under all scenarios. For five crops (irrigated grape and wheat and rainfed grape, pasture, and wheat), the costs outweigh the benefits under all scenarios. For certain crops, such as irrigated and rainfed cotton, benefits outweigh costs for this option under the High Impact Scenario only. As indicated by the dark green shading, the B-C ratios are highest for potato under the Medium and High Impact Scenarios.

These results suggest that rehabilitating irrigation infrastructure in Azerbaijan's irrigated agricultural region is not as favorable an option as optimizing the application of irrigation water. However, the B-C analysis does not take into consideration all of the benefits associated with rehabilitating infrastructural systems. As shown in figure 6.1, irrigation improves the mean crop yield, a factor captured in the study's analysis, but it also reduces risks associated with crop yield variability. The figure shows that the variation in yields of irrigated potato and tomato is much lower than that for rainfed potato and tomato. Therefore, adaptation measures featuring irrigation improvements may be favored by some farmers despite their relatively high cost. As a result, it may be desirable to

Figure 6.1 Crop Yield Variability in the 2040s for Rainfed and Irrigated Tomato and Potato in Eastern Georgia under the Medium Impact Scenario

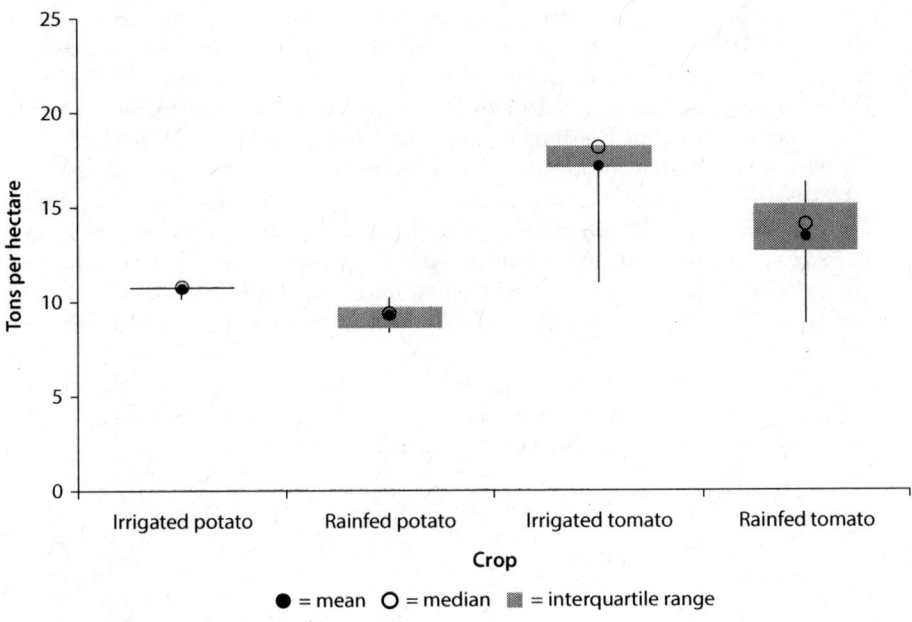

Source: World Bank data.

include both irrigation infrastructure-based measures as well as demand management measures in an adaptation plan.

In general there was significant overlap between the resulting recommendations for adaptation options for Armenia, Azerbaijan, and Georgia, both at national and agricultural region levels. Chapter 7 describes these findings in more detail and presents recommendations for developing a regional action plan for the South Caucasus region.

As described in chapter 2, the process of evaluating and prioritizing adaptation options for Armenia, Azerbaijan, and Georgia involved not only quantitative B-C modeling, but also qualitative analysis based on multiple stakeholder consultations and expert input from international and local teams. The adaptation options were ultimately prioritized from a larger menu based on the following five criteria: (1) net economic benefits, (2) qualitative expert assessment, (3) potential to aid farmers with or without climate change, otherwise referred to as "win-win" potential, (4) greenhouse gas emissions mitigation potential, and (5) evaluation by stakeholders.

The resulting national and agricultural region level options (presented in chapters 3, 4, and 5) are similar across the three countries. Chapter 7 provides insights on key regional level findings in terms of common adaptation measures and presents recommendations for developing an action plan for regional cooperation in pursuit of a climate-smart adaptation plan for the South Caucasus.

Building Resilience to Climate Change in South Caucasus Agriculture
http://dx.doi.org/10.1596/978-1-4648-0214-0

Note

1. These three countries have long been considered a regional group for the purposes of agricultural production. For example, the International Food Policy Research Institute (IFPRI), part of the CGIAR (Consultative Group on International Agricultural Research system, considers these three countries to be similar in agricultural characteristics, and has grouped them in a single Food Production Unit (FPU) for the purpose of modeling global food production, suggesting commonalities in the sector when viewed from the multicountry regional scale in agricultural markets. The FAO (Food and Agriculture Organization of the United Nations) has a 10-category "Major Farming System" agricultural land classification scheme that designates two farming system types in the region: "irrigated" and "horticulture mixed"—the only countries in the broader Eastern and Central European area made up solely of these two types. The FAO system further clarifies that there are many similarities in agricultural cropping patterns, agricultural land use, and crop suitability across the three countries. Finally, the South Caucasus area is characterized by two major transboundary river basins—the Kura and the Araks—and the vast majority of the area of this large basin is located in the three countries studied in this volume.

CHAPTER 7

Adaptation in the South Caucasus: Opportunities for a Regional Approach

As outlined in chapter 6 the risks of climate change in the South Caucasus are expected to have serious consequences for the agricultural sectors in Armenia, Azerbaijan, and Georgia. The urgency of action at the national and subnational levels cannot be overemphasized, but it is also recommended that the countries of the South Caucasus address climate change through collaboration on issues such as climate-related data sharing and crisis responses. Furthermore, the coordinated management of shared water resources would be an efficient way to address food security.

This chapter identifies the adaptation measures recommended across all three countries, both at the national and agricultural region levels, and presents a blueprint for a regional climate-smart adaptation plan based on these findings. In addition, this chapter discusses the opportunities and challenges of a regional approach in the South Caucasus.

Common Adaptation Measures at the National Level

The study identified seven national-level adaptation measures (figure 7.1), of which the following five are identical across the three countries: (1) improve farmer access to agronomic technology and information; (2) increase the quality, capacity, and reach of extension services; (3) improve farmer access to hydrometeorological capacity; (4) create a crop insurance program; and (5) improve farmer access to long-term, low-interest loans.

In general, these high-priority measures would require improving institutional capacity, increasing farmer access to information, reforming policies, and creating support programs—all essential to ensure that farm-level and private-sector actions are applied to their best advantage. To be effective, all of these initiatives require national political and financial support, so that the implementation of

Figure 7.1 National-Level Recommended Measures

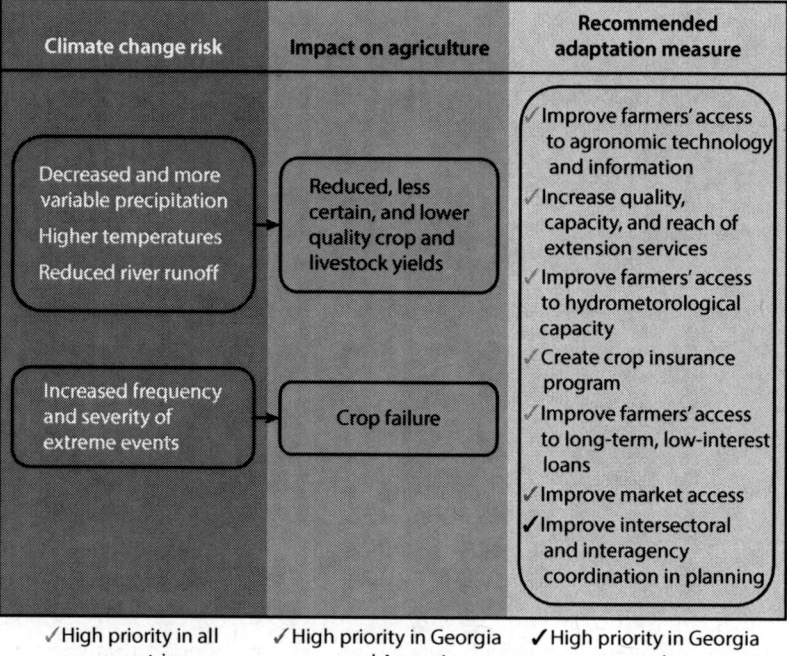

these measures could benefit by the three countries sharing their experiences and lessons learned and emulating successful programs.

Common Adaptation Measures at the Agricultural Region Level

The high-priority measures at the agricultural region level are more diverse and region-specific than those at the national level, reflecting the variety of the specific challenges farmers face. Figure 7.2 presents a synthesis of the key region-level results. Decreasing the substantial adaptation deficit of the sector is a long-term process, but several measures were identified in every agricultural region of the South Caucasus that could be undertaken immediately to strengthen the sector's adaptive capacity to climate change. As indicated in figure 7.2, the most important is to improve and optimize agronomic practices (including water and fertilizer application) and improve or change crop varieties while focusing on drought-tolerant varieties across the board.

In currently irrigated (low rainfall) areas the main challenge of coping with climate change is how to respond to irrigation water shortages that are a very likely result of warming. The most important measures are to optimize application of irrigation water, improve water availability including through rehabilitating irrigation capacity, and introduce water-efficient irrigation

Figure 7.2 Priority Measures in Agricultural Regions

High priority in all regions	High priority in irrigated regions	High priority in mountainous regions
1. Optimize agronomic practices, including fertilizer application 2. Improve crop varieties, particularly drought-tolerant crops	1. Optimize application of irrigation water 2. Improve irrigation water availability; rehabilitate irrigation capacity 3. Improve irrigation techniques 4. Rehabilitate water reservoirs	1. Adjust crop variety based on elevation 2. Research and improve livestock nutrition, management, and health 3. Construct small-volume reservoirs for water storage 4. Improve drainage infrastructure 5. Reduce erosion and practice soil conservation
Other measures	1. Establish agribusiness; assist with business palns 2. Create larger-scale farms (consolidate) 3. Establish reforestation 4. Create windbreaks	

techniques (such as drip and sprinkler irrigation). On the supply side, rehabilitation of water reservoirs that are functioning below their capacity could be considered.

The mountainous areas in the region that typically have higher rainfall and support a wider range of rainfed crops present different adaptation challenges. These areas include the subtropical region in Azerbaijan, the western mountainous region in Georgia, and the mountainous region in Armenia. In these regions climate change is likely to improve pasture productivity, so opportunities are available to farmers to enhance livestock production. In addition, in most mountainous regions changes in the variability of rainfall could lead to issues with drainage and erosion. The high-priority adaptation measures in these regions again include improving agricultural practices and livestock husbandry, constructing small-scale reservoirs (to ensure that water is available when needed even though rainfall will be more variable), improving drainage (in the highest rainfall areas), adjusting crop varieties to the respective agro-ecologies, and reducing erosion through improved land management. Specifically, in the mountainous region of Armenia, rainfall may be lower than in the other mountainous regions, suggesting that small-scale reservoirs would be even more acutely needed as an adaptation measure.

Other options that were identified as priority measures for a small number of agricultural regions were supporting agricultural markets, creating farms of suitable economies of scale through land consolidation, reforestation including agroforestry, and creating windbreaks. These measures were extensively discussed among delegations from all three countries during the Regional Knowledge Sharing Workshop in May 2013. The delegations showed strong support for both the national and agricultural region-level priorities. In addition, three new insights were shared at the workshop and confirmed among participants as key priorities in going forward.

Conservation Agriculture

First, conservation agriculture (CA) was specifically discussed as another potentially valuable agricultural adaptation strategy for the South Caucasus region. CA is a form of agricultural production based on minimum soil disturbance (no-till/direct seeding), the maintenance of soil cover, and the diversification of crops (through rotations and/or associations). The practice offers a promising agricultural land use option to produce more with less while enhancing the ecological functions of soil. The practice can offer immediate economic benefits because of reduced cultivation and machinery operations (that is, reduced costs); it also has the potential to assist in climate change adaptation by improving the resilience of cropping systems while providing greenhouse gas mitigation benefits.

CA has been introduced into Central Asia in recent years with great success. For example, Kazakhstan has 1.6 million hectares (ha) of land in CA, while research on CA in raised beds in Azerbaijan has shown good results with mixed cropping systems and has been shown to reduce erosion and increase soil organic matter (FAO 2012). A project launched by the International Center for Agricultural Research in the Dry Areas (ICARDA) set up experiments on crop diversification, bed-planting, and no-till technologies on research stations and farmers' fields in Azerbaijan (for winter wheat, maize, and sugar beet), which showed great promise. The economics of CA was tested as part of the same project in Uzbekistan, where farmers' incomes were easily doubled through double cropping even with low market prices for the alternative crops grown (CGIAR 2012).[1] A related issue raised at the Regional Knowledge Sharing Workshop concerns the need to address soil aridity. CA is one of the most effective strategies for accomplishing this goal.

Priorities for Further Study

Two other thoughts shared during the regional workshop shed light on priorities for future analysis: (1) stressing the testing of the applicability of the study's recommendations as responses to forecasted extreme events and (2) evaluating and responding to water quality concerns. The workshop discussion once again clarified the role of extreme climatic events as a challenge to the agricultural sector's productivity in the region. Off-the-shelf climate models are less appropriate for forecasting and quantifying future changes in extreme events; nevertheless several of the recommended measures are targeted at reducing effects of extreme weather, including crop insurance, hail nets, drainage infrastructure, and irrigation. In addition, a key theme at the workshop was the need to address issues of water quality and supply across multiple water-using (and polluting) sectors, including agriculture, industry, municipalities and urban areas, and hydro-power production.

Elements of a Regional Approach

The fact that many of the national-level and agricultural region-level adaptation measures are shared across Armenia, Azerbaijan, and Georgia is a compelling reason for regional cooperation in the pursuit of climate change adaptation in the South Caucasus. A key finding is that both a national enabling environment, which provides an important indicator of readiness for adopting a climate-smart plan, and context-specific and site-specific infrastructure and farmer practice measures are *both* needed to be effective.

Three main areas of opportunities for collaboration under a regional approach for adaptation to climate change among Armenia, Azerbaijan, and Georgia—to be identified among the interested countries—are the following: (1) coordinated management of water resources with due respect of riparian rights and needs; (2) collaboration on agricultural research, extension, and training; and (3) joint development of a regional network for advanced weather forecasting and climate services. In addition, as a prerequisite to taking action and in particular to gaining funding for national and regional efforts through climate finance channels, each country needs to individually demonstrate a clear readiness to develop climate-smart planning for the agriculture sector. Climate-smart planning in this region might have the following characteristics: (1) coordinated management of water resources between any of the South Caucasus countries, (2) sharing of agricultural research and extension approaches, and (3) enhanced weather forecasting provision to farmers.

Multicountry Water Resources Management

Coordinated management of water resources among the three countries—with due respect of riparian rights and any individual country's needs—has the potential to greatly reduce impacts of climate change to agricultural sector and increase shared benefits. However, regional water management planning neces-sarily requires high-level policy decisions in consideration of non-agricultural water users, including hydropower, municipal/urban water supply, and indus-trial users, as well as sustaining ecological flows and flood control. What's more, climate change may exacerbate regional pressures and conflicts, and national adaptation strategies that ignore neighboring country strategies risk ineffective outcomes.

Coordinated water management has the potential to provide three key benefits:

- Pursuit of transboundary integrated water resource management for hydro-power development and water management can create new opportunities for storage and hydropower development. Integrated water resource management provides an opportunity to optimize water use across all demand categories throughout the Kura-Araks basin, with benefits to agriculture and hydropower that could accrue to one or more riparian countries or the region as a whole. For example, increased water storage is generally constructed in higher elevation areas, where the natural terrain can be most efficiently exploited to create reservoirs and where steep slopes create greater potential for hydropower generation at the reservoir outlet. An added benefit is that the cooler higher elevation areas yield less evaporative losses from the reservoirs. A well-structured, multinational water management system could open possibilities for the higher elevation countries (mainly Armenia and parts of Georgia) to develop these storage assets and sell water and hydropower throughout the region, in exchange for other trade considerations. Co-managing the reservoirs as part of an integrated river basin system could provide multifaceted opportunities for riparians.

- Regional water quality management and monitoring would provide economic and environmental benefits across the basins. Interventions such as improved on-farm drainage and fertilizer and pesticide management by an upstream riparian would have considerable downstream benefits, such as increased production at reduced costs and improved water quality. Protection of riverine aquatic ecosystems, and the Black and Caspian Seas, will require collaboration among the riparians, with the payoff including better water quality for all uses.

- Collaboration on regional adaptation strategies is needed to maximize the shared benefits of adaptation and avoid conflicts between countries. For example, if an upstream country selects increased irrigation as its adaptation preference, its increase in off-take for storage reservoir investments and expanded irrigation could leave little or no water for downstream users. A regional adaptation approach could help avoid such outcomes. In addition, increasing irrigation efficiency in any upstream riparian will provide economic benefits to local communities as well as greater water availability downstream.

Shared Agricultural Research and Extension Approaches

Similar climate, land, ecology, and crop suitability suggest that similar crop varieties are likely to be well suited to neighboring agricultural regions, providing opportunities to share costs and benefits of research and outreach. Climate change will alter crop suitability conditions, suggesting that neighboring countries could share outputs related to research on new varieties that are adapted for the forecasted climate changes of the region—which likely will be hotter overall, drier in the lowlands, and wetter in the higher altitude areas.

In the case of new crop varieties, the Caucasus countries individually do not yet constitute a seed market large enough to induce significant private sector research effort. Significant economies of scale nonetheless exist for collaborative agricultural research between any of the countries of the region. The Consultative Group on International Agricultural Research (CGIAR) system—with CIMMYT (the International Maize and Wheat Improvement Center) already a regional partner in wheat and barley variety development—could be envisaged as a technical partner that assists to integrate respective national agricultural research programs. Countries could agree to undertake research on common agronomic issues including water-saving technologies, and pest tolerance and crop variety research on a single crop or crop groups (for example, cereals, fruit trees, vegetables). The results could be shared across the three countries through country-level extension programs based on the research. Uncoordinated research, on the other hand, risks duplication and waste of a limited research budget.

Regional Level Weather Forecasting

Enhanced weather forecasting provision to farmers could be usefully pursued at the regional level. Distribution of accurate and timely local weather information to farmers is an example of a climate adaptation service that could be effectively coordinated across the three countries, particularly to ensure optimal weather forecast sharing in cross-border and adjacent agricultural regions.

More detailed estimates of the benefits of these suggested collaboration topics among the three countries could be generated using the tools available for this the study, but such regional analyses were outside of this study's scope. For example, the Water Evaluation and Planning System (WEAP) water balance model is specifically designed for multi-sector planning, and the hydrology and basic spatial schematics were developed at the regional scale. Unfortunately, the tool has only been applied within the current study to assess irrigation measures at the agricultural region scale. Pursuit of the full set of opportunities just outlined, including enhancements to the system to evaluate water quality goals, would require additional research and data collection on the hydropower and other water user sectors. On the other hand, some of this work could be pursued by local counterparts once the tool is transferred.

Establishing a Stronger Regional Presence

In addition to this World Bank-sponsored study, a number of existing, ongoing efforts may provide continuing opportunities for regional cooperation, including the following:

- **WWF ecoregion study.** The World Wildlife Fund (WWF) coordinated a large team of researchers to develop *An Ecoregional Conservation Plan for the Caucasus*. The second edition of this report (WWF 2006) documents the efforts of dozens of experts and contributors in six countries in the region to identify sensitive ecosystems, biomes, flora, fauna, and priority species as inputs to an overall long-term vision for biodiversity planning and conservation. It appears that WWF remains interested in using the results of this work in future regional ecological resource preservation and planning efforts.

- **EU project: Trans-boundary river management phase III for the Kura River basin—Armenia, Azerbaijan, and Georgia.** In January 2012, the European Commission initiated work on a 12-month project to improve water quality in the Kura River basin through transboundary cooperation and implementation of the "river basin management approach" (Pichugin 2012). The project supports development of a common approach to water quality monitoring and assessment based on the European Union (EU) Water Framework Directive (WFD) methodologies, and it enhances technical capacities of environmental authorities and monitoring establishments to enable them to change their policies and practices in accordance with WFD. The project involves (1) development of a common approach to water quality assessment based on existing data and EU WFD methodology; (2) capacity-building and training on policy and technical guidelines to facilitate adoption of the common approach to water quality assessment; (3) joint field surveys in transboundary pilot basins, including water sampling and analysis; and (4) coordination of water projects in the South Caucasus region implemented by EU and other international agencies. The work builds on the results and experiences of the successful EU Kura Phase II project completed in December 2011 with the cooperation and approval of the European Commission and the beneficiary countries of Armenia, Azerbaijan, and Georgia (Pichugin 2012).

- **REC Caucasus agro-biodiversity study.** The Regional Environmental Centre for the Caucasus (REC Caucasus), a nonprofit local organization, recently embarked on the EU-funded multicountry effort, "Identification and Implementation of Adaptation Response to Climate Change Impact for Conservation and Sustainable Use of Agro-Biodiversity in Arid and Semi-arid Ecosystems of South Caucasus." (REC Caucasus 2011). The geographic scope of the effort is identical to this study, and REC Caucasus is currently developing maps at the district level of areas where the project will focus. Joint recommendations for preservation of biodiversity across the region are an expected outcome of the study.

- **UNDP-sponsored climate impact study.** The recently completed *Regional Climate Change Impact Study for the South Caucasus Region* (UNDP 2011) provides relevant information for the purposes of mapping ecoregions, as well as climate impacts and adaptation in the transboundary river basins that characterize the region. This joint effort was coordinated by the United Nations Development Programme (UNDP) and was conducted by a consortium of the Ministry of Nature Protection of the Republic of Armenia,

Ministry of Ecology and Natural Resources of Azerbaijan Republic, and the Ministry of Environment of Georgia, as well as a team of local researchers. A follow-on effort is conducting additional analyses of water resources on a regional scale.

These recent efforts underline the need for a more coordinated strategy across the region. At this time, efforts by the World Bank, UNDP, EU, and WWF are pursuing related environmental management and climate adaptation goals regionally, with representation from each of the three countries. The coordination and oversight of these efforts would greatly benefit from a regional entity set up to systematically ensure consistency and lack of duplication of effort and to provide greater legitimacy of the results for each of the region's governments.

An initial step in this effort would involve collaborative monitoring, evaluation, and reporting. A potential model for this is the Interstate Commission for Water Coordination (ICWC), centered in Uzbekistan, which coordinates water-related data collection and reporting for a substantial portion of Central Asia, including Kazakhstan, Kyrgyz Republic, Tajikistan, and Uzbekistan. The ICWC includes two Basin Water Organizations (BWOs), for the Syr Darya and Amu Darya basins, which are executing agencies of the overarching international ICWC authority. The ICWC data portal includes water intake volumes, flow, and water demand data for each of the two major BWOs (http:// www .cawater-info.net). In addition, river basin schematics and maps are also available at the portal.

For a short period, ICWC also included an adjunct network of Eastern Europe, Caucasus, and Central Asia (EECCA) water-management organizations, which included Azerbaijan and Georgia as members. The network was established pursuant to discussions in 2007 to exchange views, experiences, and information on various aspects of water-management activity, but at press time it appears to have been deactivated. In the future, the mission for the Caucasus countries could be expanded to share data and serve as a periodic forum to pursuing joint water management needs (United Nations 2009).

Challenges for Regional Resource Needs

Pursuing climate adaptation goals in the agriculture sector through regional collaboration requires three types of resources: human, financial, and information.

Human Capital

Human capital in the region resides in domestic institutions, international finance institutions (IFIs), nongovernmental organizations (NGOs), and international organizations such as CGIAR, World Meteorological Organization (WMO), universities, and partnerships established through cooperative efforts, such as in this study. In all three countries, research and extension have a strong tradition but are not yet oriented toward adapting to current and forecasted climate challenges. Unfortunately this human capital is not coordinated across countries, leading to unnecessary and costly duplication. Thus, the commonalities in the region of climate, agro-ecosystem types, and shared water resources are not currently exploited efficiently and effectively.

Financial Capital

Financial capital should be considered in the context of current priorities, options for national government financing, and external financing sources that can be applied to address gaps in financing. In most developing countries, climate adaptation is not mainstreamed with the result that external sources are a main source of finance. Therefore the next section focuses on these sources as important to pursuit of climate adaptation initiatives. However a more thorough analysis of financial capital needs is needed and should also consider the extent to which the recommended actions align with current agriculture sector priorities. One such example is providing better access for farmers to high-quality hydrometeorological forecasts. Therefore, such an analysis would consider the extent to which existing national financial capabilities could be applied to achieve the goal of enhanced climate resilience in the sector.

External finance now flows from donor and local resources, including development partners, and will flow from the new climate finance options that are coming on-line. New opportunities in climate finance may provide a strong incentive to recognize the benefits of regional economies of scale, with an initial focus on national governments. Climate finance can be sourced through private, public, or joint private-public sources and are often channeled through some type of intermediary, such as international and bilateral financial institutions (IFIs and BFIs) and development agencies. Experience to date shows that preparation of an integrated, high-quality strategic plan based on sound analysis and deep understanding of the challenges, opportunities, and potential trade-offs are critical steps. The result is a strategic programming for investments capturing green, clean, resilient, and inclusive growth options. Such planning processes require improved awareness among the various stakeholders (for example, senior government officials in key line ministries, civil society, parliamentarians, private sector) on the need for changed development pathways.

Analytical work articulating the potential costs of current climate risks to development goals and the costs and opportunities to move to greener, climate-smart options are also important for elevating these development options into key ministries. Access to good quality information and analysis and a systematic climate risk assessment using historical and projected changes in climate, their impacts, and options for minimizing risks to development are important in overall planning of these investments. These steps will also indicate to potential donors that recipient countries are ready to access these funds.

In order for the financial resources of adaptation financing to be effective, information systems, technical and managerial capacity, and the right policies and institutions must be in place. Without such governance structures, the impacts of adaptation finance may be diminished. It is also important to note that the wide ranging capacities of institutions and actors in developing countries may result in differing outcomes from similar levels of adaptation financing.

Information Resources

Information resource needs in the South Caucasus region include quality and quantity information on the natural resource base, economic information on the efficiency of adaptation measures within and across sectors, and options for incorporating internationally available adaptation measures (such as new crop varieties).

The fulfillment of information needs depends on three sources: (1) analytic work from this study and the ongoing others discussed here, (2) qualitative information from farmers and local experts and policy makers, and (3) globally available information, ranging from pure data to academic knowledge. The last category in particular is an excellent starting point for region-scale collaboration to learn about options like climate-smart agriculture, for example, including conservation tillage. Generally, this type of information can be sourced from organizations such as FAO and CGIAR. In addition, the WMO is capable of providing both up to date information on climate and climate forecasting tools and capacity building in this area.

Developing an Action Plan

Each of the three South Caucasus countries must develop its own action plans for the priority measures the study identified at the national and agricultural region levels. These countries should also address climate change through collaboration on issues such as climate-related data sharing and crisis responses. Furthermore, the management of shared water resources would be efficient to address food security. Where knowledge, skills, or technology are lacking in one country, they often exist in other countries, thus complementing each other.

Table 7.1 can serve as a starting point for pursuing the three strategic elements of a plan for greater South Caucasus regional collaboration in adapting the agriculture sector to climate change. Although there are many challenges to achieving these objectives, fortunately there is also a wide range of existing "models" of regional-scale institutional arrangements throughout the world, encompassing the scope of regional cooperation for water resources planning, agricultural research and extension, and enhanced hydrometeorological service development and data provision.

Table 7.1 Summary of a Regional Agricultural Sector Climate Adaptation and Mitigation Plan for the South Caucasus Countries

Strategic elements	Objectives	Coverage/ stakeholders	Potential issues and barriers to overcome	Responsible authority	Existing models for collaborative efforts	First steps	Key outputs	Possible funding sources
1. Coordinated management of water resources between South Caucasus countries	• Reduce impacts of climate change to agricultural sector • Increase shared benefits, particularly for storage/ hydropower development • Maintain ecological flows and water quality	Farmers Non-agricultural water users (urban) Hydroelectric power companies Industry	Riparian rights National-level needs may conflict Water flow and quality data may be inconsistent	Initially national ministries Once established, a joint or several Basin Authority(ies)	Interstate Commission for Water Coordination (Central Asia) International Commission for the Protection of the Danube River	Establish tri-partheid working group Hold semi-annual meetings	Co-managed Basin Authority(ies) for each basin established Collaborative management capacity developed Knowledge and decision-support products disseminated and maintained Kura-Araks River Basin Management Plan	International financial institutions Other donors Domestic budgets User-pays mechanisms
2. Collaboration on agricultural research and extension	• Reduce impacts of climate change to agricultural sector • Jointly access and influence CGIAR research • Gain economies of scale in extension	National research institutes CGIAR system entities National extension services Private extension services (where applicable)	Rights to agricultural technologies Current extension may be poorly subscribed or relied upon by farmers	National ministries of agriculture and education	CGIAR components	Identify and prioritize research needs Host a joint summit meeting with national stakeholder, CGIAR representative, and funders	Research results shared across the three countries Country-level extension programs incorporate the new research results in demonstration plots and trainings National-level research better coordinated	CGIAR system (donor- and domestic budget-funded) International Fund for Agricultural Development UN Food and Agriculture Organization Domestic budgets User-pays mechanism for extension and seed products

table continues next page

Table 7.1 Summary of a Regional Agricultural Sector Climate Adaptation and Mitigation Plan for the South Caucasus Countries *(continued)*

Strategic elements	Objectives	Coverage/ stakeholders	Potential issues and barriers to overcome	Responsible authority	Existing models for collaborative efforts	First steps	Key outputs	Possible funding sources
3. Enhanced weather forecasting provision to farmers pursued at the regional level.	• Reduce impacts of climate change to agricultural sector • Expand capabilities of hydromet services for the region	National hydromet institutes Farmers WMO	Existing monitoring equipment and data collection may be inconsistent Intellectual rights to data can be complicated and limit sharing of products across countries and with farmers	National hydromet institutes in partnership with users and stakeholders at various scales	Climate Services Partnership Caribbean Institute for Meteorology and Hydrology International Research Institute for Climate and Society (e.g., see http://scalingup.iri.columbia.edu/index.html) AGRHYMET Regional Center (extreme events forecasting)	Establish a working group with representatives of hydromet to identify goals and data and information gaps Study international cooperative efforts for ideas and to clarify institutional arrangements Reach out to fundraisers and prepare a proposal for start-up activities	Distribution of accurate and timely local weather information to farmers Creation of new long-term and extreme event forecasting capabilities for regional purposes	WMO U.S. Agency for International Development Domestic budgets User-pays mechanism for some enhanced weather products and delivery modes

Note: CGIAR = Consultative Group on International Agricultural Research; hydromet = hydrometeorological; WMO = World Meteorological Organization.

Note

1. Maize was double-cropped after the winter wheat harvest in Azerbaijan and provided 4.9 tons per ha (91.25 percent) yield advantage after no-till wheat. The CA-associated bed-planting method was shown to improve yield (with a maximum wheat yield of 5.51 tons per ha) and to save seed and water (an average of 36 percent water).

References

CGIAR (Consultative Group on International Agricultural Research). 2012. *CGIAR Regional Program for Sustainable Agricultural Development in Central Asia and the Caucasus: Annual Report 2011–2012*. Montpellier, France: CGIAR (accessed December 9, 2013), http://cac-program.org/files/15scm_cac_annual_report_en.pdf.

FAO (Food and Agriculture Organization of the United Nations). 2012. *Conservation Agriculture in Central Asia: Status, Policy, Institutional Support, and Strategic Framework for its Promotion*. Rome: FAO (accessed December 9, 2013), http://www.fao.org/docrep/017/aq278e/aq278e.pdf.

Pichugin, A. 2012. "ENPI Project: Trans-Boundary River Management for the Kura River Basin, Phase III: Armenia, Georgia, Azerbaijan." Paper presented at National Policy Dialogue Meeting, Yerevan, April 11. http://www.unece.org/fileadmin/DAM/env/water/npd/ENPI_project_trans-boundary_river_management_for_the_Kura_river_basin_Phase_III.pdf.

REC Caucasus (Regional Environmental Centre for the Caucasus). 2011. "Identification and Implementation of Adaptation Response to Climate Change Impact for Conservation and Sustainable Use of Agro-Biodiversity in Arid and Semi-arid Ecosystems of South Caucasus." http://www.rec-caucasus.org/profile.php?id=1323939165&lang=en.

UNDP (United Nations Development Programme). 2011. *Regional Climate Change Impacts Study for the South Caucasus Region*. Tblisi, Georgia: Environment and Security (ENVSEC) Initiative and UNDP (accessed October 17, 2013), http://www.envsec.org/publications/cc_report.pdf.

United Nations. 2009. *Capacity for Water Cooperation in Eastern Europe, Caucasus and Central Asia: River Basin Commissions and Other Institutions for Transboundary Water Cooperation*. Economic Commission for Europe: Convention on the Protection and Use of Transboundary Watercourses and International Lakes, New York and Geneva (accessed October 16, 2013), http://www.unece.org/fileadmin/dam/env/water/documents/cwc%20publication%20joint%20bodies.pdf.

WWF (World Wildlife Fund). 2006. *An Ecoregional Conservation Plan for the Caucasus*. 2nd ed. Tbilisi: WWF. http://assets.panda.org/downloads/ecp_second_edition.pdf.